Signs of Life

Signs of Life

A Semantic Critique of
Evolutionary Theory

Ashish Dalela

SHABDA
PRESS

Signs of Life—A Semantic Critique of Evolutionary Theory
by Ashish Dalela
www.ashishdalela.com

Published by Shabda Press
www.shabdapress.net
ISBN 978-93-85384-00-4

Dedicated to all atheists. Listening to your arguments about what science is telling us about the nature of life, I was inspired to write this book.

Man's mind, once stretched by a new idea, never regains its original dimensions.

—Oliver Wendell Holmes, Sr.

Contents

List of Figures

Preface

The key idea that I wish to describe in this book is that the theory of biological evolution is based on concepts that are inconsistent with the notions of matter in modern physics, the problems of meaning in mathematics, what randomness entails in computing theory, and what game theoretic and ecological perspectives tell us about the natural world. Many of the ideas Darwin formed in his theory of evolution were based on notions of space-time, matter, and causality used in classical physics. Deep connections between physics and chemistry had not been made then, and consequently the problems of physics and their implications for chemistry did not have a bearing on biological theories. Mathematics was supposed to be logically complete and consistent, and its incompleteness and paradoxes had not been shown. Computing theory did not exist and its problems did not have an impact on biology. Game theory had not yet been invented, and its implications for competitive behaviors were not known. Over the decades after Darwin much has happened in other fields outside biology, but, unfortunately, the problems of causality, motion, completeness, complexity, non-linearity, and meaning in these other fields have not had any serious impact on biology. Biology still lives in a relative time-warp of philosophical presuppositions that were acceptable in Darwin's time but are now false.

Let's take for example the conceptual revolutions in physics. The ideas of causality, motion, and space-time have undergone a dramatic revision in quantum theory but they have had little impact on biology. Biologists still think of changes as being caused by forces when, in fact, quantum theory states that a closed system is always in a stationary state. Biologists believe that molecules are real things, when quantum physics tells us that the same ensemble of particles can be divided into many different sets of molecules. Biologists still think that a living being is built up of thing-in-itself particles, ignoring new ideas such as entanglement and non-locality. The unresolved measurement problem in quantum physics—which entails that objects are not real until a measurement is performed—has had literally no impact on biology; biologists continue to think that there is in fact a reality which exists without observation.

There are considerable conceptual problems when ideas in physics have to be extended to living beings. For instance, the brain can hold ideas about the world but physical objects can only indicate their own properties. But this problem is generally ignored because physicists study non-referential objects, and biologists extend this view of nature even to living beings, which are referential.

The problem of reference appears in mathematics and computing theory as the incompleteness and incomputability of all number systems, although the connection between incompleteness, incomputability and reference is generally not well-understood. Again, those scientists working outside mathematics take a pragmatic approach to mathematics: numbers are useful for calculations but what numbers are is not relevant to other areas of science. In particular, problems of mathematics or computing are not problems of physics or biology because numbers are not physical. However, these problems become very important when a physical theory has to explain the ability to process meanings in the brain. Now, numbers have to be treated as concepts rather than quantities. A true conceptual treatment of numbers has never been successful because mathematicians have tried to derive number concepts from objects. This in turn entails that no formal system (in the sense defined currently) can ever contain meanings. If biological systems are described using such mathematics, they too could not have meanings. This is not a question of biological complexity; it is rather a problem of not possessing a mathematical formalism in which to describe a physical system that can represent meanings about the world.

If classical notions of causality, motion, and determinism fail so noticeably in physics, then they must also fail in chemistry and biology. If the problems of meaning have not been solved in mathematics and computing theory, then surely they could not be solved in chemistry and biology that simply inherit the logical thinking. If the creation of sub-atomic particles cannot be predicted in physics, then the formation of bonds and chemical molecules also cannot be predicted. Evolutionists seem oblivious to these challenges. They suppose that all fundamental issues about the nature of atoms, space-time, and causality have been settled, and they now wish to settle the issue of how molecules form living beings. They presume that physical systems can encode meanings, when this is known to be logically impossible in a physical system. The paradox here is not that living bodies consist of molecules. The paradox is that the theory of molecules themselves is incomplete. Could it be that when the physical theories are completed, they would also be able to explain the meaning forming capabilities in living beings? Could it be that this ability would require a substantial revision to logic and mathematics, since meanings cannot be encoded in physical objects?

The problems of physics and mathematics are not problems of nature. It is very likely that causality in nature, and logic and numbers are quite different from how we have currently conceived them. To build a causally complete theory, new notions about causality are needed. To incorporate meanings in the material world, new notions about logic and numbers are necessary. These new ideas about matter, logic, and numbers would not just change our view of physics and mathematics, but also our ideas about living beings. Without their solution, biology would be incomplete, too.

This book extends my earlier work in physics, mathematics, and computation into biology. The first chapter in the book surveys some basic intuitions about the evolution of ecosystems, organizations, and societies and what these forms of evolution may have to do with biological evolution. The second chapter describes how evolutionary theory depends on ideas about matter and causality that are not true in modern physics; they were of course true in classical physics. The chapter also discusses the problem of meaning as it appears in mathematics, computing theory, and linguistics, entailing a fundamentally different view of nature, which is incompatible with the materialist-reductionist ideology in biology. The chapter hints towards potential pathways that can be used to bridge the gaps. The third chapter describes a newer approach to the problems in quantum theory, mathematics, and computing, and summarizes my prior work on the role of meaning within science. The central theme of the chapter is how the mind-body dualism that forms the bedrock of modern science (and yet creates many types of foundational problems) can be solved. The fourth chapter shows the relevance of this semantic view to biology and what the solution to the mind-body problem will entail for biology. The fifth and final chapter contrasts these ideas with other alternatives to evolution.

I call this way of looking at matter the *semantic view* because it treats material objects as symbols of meaning rather than meaningless things. If the semantic view resolves foundational problems in physics, mathematics and computing, I find it a more compelling approach to biology than thinking of meanings as random events.

In other areas of science, randomness is considered a problem, not a feature of reality. For example, physics has probability in quantum theory, relativity has indeterminism associated with gauge transforms, chemistry has uncertainty in molecular states, mathematics has incompleteness, and computing theory has incomputable problems. In no field of science, except biology, are chance, uncertainty, probability, and indeterminism welcome features of science, let alone reality. In each field, there are attempts to solve problems of indeterminism and incompleteness. Although many such attempts have failed in the past, the efforts continue. Biology, therefore, has the dubious distinction of celebrating what other areas of science will consider a problem worth solving. By suggesting that information originates in random events, we are overlooking potential sources of causality in nature. If physics, mathematics, and computing were to address the problems of indeterminism and incompleteness, there would be no reason to suppose that biology will have any theoretical room to postulate random chance events.

Shifts in fundamental theories of nature will therefore have dramatic effects on the mainstay in biology—evolution. This book describes how such shifts are implied by current physical theories but require such a fundamental shift in thinking that they would not be brought about without a dramatic change in our notions of matter, space-time, and causality. The book discusses what the new notions could be. Structure and function in biology, in this new view, are not products of random chance events. They are rather systematic outcomes of the same process of information creation by which we create books, music, art, and science. A biological species is the representation of an idea. Like a book encodes ideas into matter, the body of the living being similarly encodes ideas into structure.

The evolution of species can now be modeled by the same laws that govern the evolution of ideas. As ideologies come into favor and then fall out of favor, species are also created and modified. The central question now is: Can we predict biological evolution using this model? I will argue that we can. The ability to predict evolution is a distinct advantage, because the current theory of biological evolution does not predict it. The current theory speaks about random mutations at the molecular level, which are so hopelessly complex that physicists have found it impossible to solve equations for anything but the simplest of the molecules, even with many approximations. Only an atomic theory capable of predicting macroscopic changes can predict biological evolution. The macroscopic theory of quantum phenomena also requires a new kind of mathematics. A discussion of the problems in physics and mathematics and their evolution is thus relevant to the biological questions of evolution.

Similarly, the problems of meaning in computing and mathematics constrain the form of the theory that can describe meanings in living beings. The ability to perceive and conceive, to have intentions and knowledge, are issues central to an understanding of life. These are largely neglected in science and, by implication, in biology. The dominant dogma in biology holds that these reduce to the interactions amongst molecules. However, how such interactions create quality, contextuality, and intentionality—which are central features of any experience—remains a deeply mystifying problem. Traditionally, if we cannot reduce a feature of reality to known features, we postulate that the unexplained feature is fundamental in nature. However, the postulates that arise from the need to explain meaning and mind are so antithetical to the current materialist dogmas that it has become nearly impossible to talk about these ideas without the discussion being labeled as pseudo-science.

This book aims to show how the ideas of mind and meaning can be scientific not just in the context of biology but also in physics, mathematics, and computing, which are generally seen as being beyond reproach as far as mind and meaning are concerned. The book will argue that nature is not governed by separate theories of evolution that prevail at different levels of complexity. Rather, nature can and should be explained by a single theory that governs both inanimate matter and living bodies. However, a theory that can truly bridge the current divide between living and inanimate matter will also paint a view of matter, space-time, and causality quite different than that prevalent in the current theories of inanimate matter. Specifically, ideas about meaning and intentionality need to be brought into the discussions about numbers, objects, causality, and laws.

Science has made tremendous progress in the past by keeping the mind outside of science. What then tells us that the approach needs to be changed now? The answer to this question is that the problems of inconsistency, indeterminism, incompleteness, and irreversibility force us to reexamine the eviction of mind from science. The reconciliation between animate and inanimate matter under the same theory cannot be achieved until mind is reintroduced in science in a way that is useful even to inanimate descriptions. This shift in science probably represents a greater change in our thinking than the earlier one which separated mind from matter. I don't expect this journey to be easy, but I do expect it to be fruitful.

1

The Big Picture of Evolution

I believe that mechanism in biology is a prejudice of our time which will be disproved. In this case, one disproof, in my opinion, will consist in a mathematical theorem to the effect that the formation within geological time of a human body by the laws of physics (or any other laws of similar nature), starting from a random distribution of the elementary particles and the field, is as unlikely as the separation by chance of the atmosphere into its components.

—*Kurt Gödel*

Setting the Scene

The modern molecular theory of evolution comprises three main ideas. First, that species are created and destroyed; the species that exist today have evolved from other species in the past, which may now be extinct. Second, that the evolution of species does not have a direction and the species are not 'headed' towards an ultimate goal, such as towards some utopian perfection. Third, that the *mechanism* of evolution is random chance events followed by natural selection. Random chance events imply that evolution takes place without volition or free will. Natural selection means that if a mutation improves the chances of survival for the mutating species in some given environment, then the species lives on while others perish.

The modern theory of evolution differs from its Darwinian ancestor in the *mechanism* by which evolution takes place. In Darwinian theory, too, there was evolution from ancestral species and the evolution did not have a direction or ultimate goal. However, Darwin's theory spoke about the role that the environment plays in the acquisition of new traits which are then passed on to the offspring. This idea is called *pangenesis* and indicated Darwin's supposition that traits in living beings are enhanced or reduced based on "use or disuse." This theory is now known to be false, as a carpenter's son is not necessarily a better carpenter. Darwin's pangenesis was an extension of Lamarck's idea that an organism passes on the acquired traits. The mechanism of evolution in the modern evolutionary is instead based on Mendel's work on genetics. The modern evolutionary theory claims that molecules can randomly mutate, and when the mutation occurs in the genetic material—which is passed to an offspring—then a new biological trait has been randomly created. If this trait helps the offspring survive better, then random mutations can be used to explain the evolution and diversification of species.

There is considerable evidence today on the existence of genetic mutations. There is also no physical or chemical theory that explains or predicts which of these genetic mutations will occur. Without a theory that can predict these mutations, it seems compelling to assume that these mutations are in fact random. Of course, a theory that has randomness as its foundation is not a scientific theory in the conventional sense of a cause-and-effect model. If the cause is random then the effects resulting from this cause must also be random. How could we use such a theory to make predictions?

Indeed, Karl Popper—a philosopher of science—at one time argued that evolution is not a scientific theory because it does not offer predictions which can be potentially falsified. According to Popper, all scientific theories must in principle be falsifiable, which would happen if the predictions of the theory were proved to be contrary to the facts. Since evolutionary theory does not make predictions which can potentially be falsified, it could not be a scientific theory in sense of a theory that can potentially be falsified. Popper later revised his stance and said that evolution is a scientific theory although it is very difficult to make predictions with it. This is indeed reflected in the fact that evolutionary theory does not offer predictions about how the current species may evolve in the future.

Most biologists today use evolution to 'make sense' of the historical data, not so much to make predictions. One such type of data is the genetic similarity between species. The human and the chimpanzee DNA are 99% similar and this could be explained if we assume a common ancestor from which both humans and chimps have evolved. Evolution also helps explain why some species may become extinct if they are not able to adapt to the environment, which is indicated by the study of fossils and their carbon dating.

And yet evolution involves some leaps of logic. The idea that there is genetic similarity between several species is different from the idea that this similarity is *caused* by evolution. Genetic similarity between the DNA of two species does not conclusively prove that these species had a common biological ancestor, although a common ancestor is a possible explanation of the genetic similarity. Likewise, the carbon dating of fossils suggests that species appeared in a specific order although the order of their appearance does not conclusively demonstrate that one species *evolved* from another.

In any field of science the data underdetermines the theory. Generally, there are many possible explanations for the same data, and the theory that makes fewer assumptions, involves fewer leaps of logic, and explains a wider variety of phenomena is chosen over others. Evolutionists generally focus on the fact that evolutionary theory is optimal because it makes fewer assumptions. On the face of it, the theory only requires two ideas—random mutation and natural selection—both of which are empirically confirmed. So, it would seem that the theory is based on ideas that are already empirically validated and therefore seem to be good assumptions.

However, there are also serious questions about whether random mutations followed by natural selection can generate the kind of complexity that exists in biological entities. This issue is often debated in the context of biological species that seem to require transitional forms to arrive at the present state, although no transitional form fossils are observed. This is also widely discussed in the Intelligent Design (ID) community based on the fact that to even select, species must have a functional form, which begs the question of how these forms were originally created. Honestly, I am not an expert in all the evidence that exists for and against evolution, and I will therefore steer clear of debating that evidence and what it means for evolution, one way or another. I will rather be preoccupied with some theoretical issues in the evolutionary theory itself.

These theoretical issues may be called meta-biological concerns, for lack of a better term. Most people who hear the term meta-biology for the first time may think of metaphysics, which is a branch of philosophy that deals with the kinds of things that exist or could exist. If meta-biology were just like metaphysics, then it would be a branch of philosophy that describes the nature of life. But that is not what I am aiming for. The term 'meta' means that which lies outside or beyond, and by meta-biology I mean issues that lie outside biology—in physics, mathematics, computing, and game theory—and which have something significant to say about biology. I believe that there are some very important theoretical constraints that emerge when well-established ideas in computing, physics, mathematics, and game theory are brought to bear on biology.

For instance, I will discuss arguments from Gödel's incompleteness theorem in mathematics and Turing's halting problem in computing to show why the theory of evolution is either inconsistent or incomplete. I will show that the theory of evolution has a similar logical *form* as the problems that led to incompleteness in mathematics and incomputability in computing theory. Thus, if the theory uses both random mutations and natural selection, then it is inconsistent and if it uses only one of them then it is incomplete.

I will also discuss problems in quantum physics to show that the idea of randomness in biology has a basis in quantum statistics and this puts the idea of random mutations on the same level as the unsolved quantum measurement problem in physics. Quantum measurement is an unsolved problem because if current quantum theory were indeed a final theory of matter then the everyday world—including living beings—could not exist as stable objects. Quantum statistics restricts what can safely be said about molecules, chemical bonding, and hence about molecular mutations, although most chemists don't take these facts about quantum theory very seriously. Chemistry converts the quantum indeterminism into a classical determinism, thereby ignoring the problems that arise from a profound appreciation and understanding of these issues.

The problems of causality in quantum theory leave gaps in our understanding of how energy and electrons flow in a biological system—for instance carrying nervous impulses to the brain. Many things that biologists assume—such as the existence of molecules, motion, and force-based causality—are ideas that were true in classical physics but have undergone a radical denial in quantum theory. New notions about objects, change, and causality have, however, not emerged so far, and most physicists continue to use the classical ideas as a pragmatic working hypothesis in science. This is especially true of chemistry, biochemistry, and biology, where ideas drawn from classical physics—such as motion, force, and things—continue to be used in the same way as they were prior to quantum theory.

Of course, this does not imply that there isn't matter, change, and causality in nature although it does imply that the classical notions about matter, causality, and change are incorrect. What does the failure of classical physical concepts in quantum theory entail for biology? I will show that quantum problems unseat many classical ideas about matter, causality, and change that biologists use, and that disruption in turn entails radical revisions to the ideas about the mechanism in evolution, in ways we don't yet acknowledge.

There are also issues arising from representational and computational capabilities in the brain, given that Gödel's incompleteness and Turing's halting problem show that there cannot be physical systems that can consistently represent and compute meanings. How can the physical processes of random mutation and natural selection create brains with meanings when there cannot be a complete and consistent theory of meaning in mathematics and computing? Obviously, semantics requires us to incorporate meanings in mathematics and then in physics, before biology can use them. That too changes the way we have to look at material objects, how these objects make up living beings, and what it means for evolution.

But of even greater importance is the fact that even if random mutations and natural selection were theoretically sound ideas (i.e. that they were based on sound fundamental principles in physics, and if they could be mathematically consistent and complete) they would still not explain the creation of new species. This conclusion follows from the application of non-linear dynamical principles to biological ecosystems, because non-linear dynamics tells us that an ecosystem must either oscillate violently (called chaos) or it must sit in relatively stable states (called attractors). If the ecosystem oscillates violently, it will kill most of the species. If, however, the ecosystem sits in stable states, it will not create any new species. I will discuss how current biology treats the biological ecosystem as a linear system which evolves gradually, like a slow moving ball. In a linear system, causes flow in one-direction rather than returning to act on the original cause to cause a system-wide oscillation. Assumptions about linearity lead to the notion that new species can be created, although a non-linear system forbids this viewpoint.

Much of the discussion in biology today happens without a deep appreciation of these problems. We sort of assume that issues of causality and completeness have already been settled in physics and chemistry, and all that remains to be said is how living beings emerged from molecules. We also assume that minds can be reduced to matter although there are some fundamental logical and mathematical problems in carrying out that reduction, in principle.

Most of the current oppositions to the theory of evolution delve far more into the empirical evidence on whether the theory is consistent with the observations, but seldom on the conceptual and logical issues in the theory itself. This book hopes to fill that gap in the current discourse about evolution. I will not expend effort in debating the empirical evidence, but outline the form of the theory that can conceptually and theoretically be consistent with what is empirically and theoretically known in physics, mathematics, computing, and what we intuitively know to be the problems of encoding meaning in matter. While brains are part of the meaning problem, the question of meaning is not tied to the brains. Rather, the problem of meaning can also be discussed in the context of semantic computing machines, the issue of semantics in mathematics, and the revisions to physical theories needed to encode meaning within matter.

It is noteworthy that science is a combination of theory and observations. Unless a new theory explains the old and new observations in an encompassing manner, the current theory continues as the pragmatic and heuristic instrument for organizing data. While there is mounting evidence that the current theory of evolution does not explain all the available data, it is also important to recognize the theoretical issues in the current views about evolution. This book hopes to bridge the gap between what is well-known in areas of science outside biology—computing, mathematics, game theory, and physics—to what biology needs in order to explain the observed biological diversity. By stepping outside the traditional debates about the evidence that stands for and against evolution, I hope to show many fundamental problems in the theory itself.

The Alternative to Molecular Evolution

However, discussing the flaws of evolution is not the main goal of this book. I will discuss the flaws only to illustrate the need for an alternative view. The rest of the book is about what the alternative can be. I believe there is an alternative theoretical explanation of biological similarity that employs some new fundamental properties in nature. I will argue that these properties are *structure* and *function*. If structure and function are fundamental properties, and different species embody these properties in different ways, then what is called mutation in current evolutionary theory would be studied as the change in structure and function. Like current physical theories describe the evolution of an object's state—because we believe that aspects of that state such as position and momentum are fundamental—there can be new causal physical theories that describe the evolution of structure and function. Just as current quantum physics explains all chemical elements based on six quarks and six leptons, similarly, we could explain many different structures and functions based on a finite number of structures and functions.

In a theory that treats structure and function as fundamental properties, we could be asking the following interesting questions. Which are the elementary structures and functions from which other structures and functions can be created by combination like atoms are created by combining sub-atomic particles? What causes the structure and function to combine, and what kinds of laws should govern this combination? What is the relation between a structure and a function, given that many structures can enact the same function and a structure can be used for many functions?

Answers to these questions will require a theory about new types of logical entities, which can be described mathematically, and can be used for making definitive predictions. Like chemical element groups and the similarity between different chemical elements are an epiphenomenon of current quantum physics, the biological similarity between species can be an epiphenomenon of such a theory because similarities between species can be reduced to the similarities between structure and function in species. Like Neon, Argon, and Krypton have similar properties of inertness because of electron orbital occupancy, it would be possible to speak about the similarity of different species based on fundamental structures and functions. The laws of nature in such a theory will speak about why some structures and functions are fundamental, how these elementary structures and functions are created, the laws by which these types of entities combine, and the causality that dictates their evolution. Evolutionary theory will now be replaced by a theory that answers the basic questions about structures and functions.

Of course, a fundamental theory about structures and functions hinges on the possibility that structure and function are indeed *fundamental* rather than *emergent* properties in nature. In current biology, structure and function are viewed as emergent properties. Therefore, before we seek to formulate theories about structures and functions, we need to be convinced that these properties are indeed fundamental. That insight itself requires us to step outside the conventional boundaries of current biology. Quite specifically, the idea that structure and function are fundamental (rather than emergent) properties in nature requires the following three steps.

- We need to understand why all theories that aim to describe nature in terms of individual particles or objects are incomplete. Such object theories include classical mechanics, statistical mechanics, quantum theory, and general relativity. I will describe why each of these theories is now known to be incomplete. Unless we are convinced that the reduction of nature to objects is incomplete, structures and functions will remain emergent properties and there would be no need to seek alternatives.

- We need to recognize that the incompleteness of object-based theories arises when these theories attempt to describe *ensembles* or *collections* rather than individual objects. This realization helps us find the root cause of the incompleteness, namely that a theory becomes incomplete when it tries to reduce a collection to independent parts. If all theories that reduce collections to independent parts are incomplete, it follows that collections should not and could not be reduced to independent parts.

- We can now postulate that there must be some properties in collections which cannot be reduced to the properties of the parts in them. Clearly, every ensemble or collection aggregates objects in structures. If these structures cannot be reduced to the constituent objects, then a new kind of theory of structures must be devised. This theory will also inherit some properties of object-based theories. For instance, we could ask which structures are fundamental, how is the state of a structure described, and the laws by which these structures interact and evolve.

Current physical theories reduce all structures to the positions of independent particles that participate in that structure. The idea is that the particles are real and they aggregate to form structures. But how does the aggregation occur? Evolutionary theory argues that these aggregations are randomly created and then naturally selected based on a mutual 'fit' between structures. The key underlying assumption in this view is that a particle that participates in a structure is unchanged by that participation itself. The same particle could have been part of another structure, and it would not materially change anything in that particle. This assumption in the reduction of structures to parts is false, as I will discuss subsequently.

For instance, an electron in two different molecules is not the same electron because quantum theory prescribes different eigenfunctions to describe that electron. In fact, even different electrons within a given atom are not identical; the electrons that form the *s* orbital are different than those that form the *p* orbitals. There is nothing in physics that allows us to assert that any two electrons in a quantum system are identical and 'could have' been in another state. Indeed, we cannot distinguish between the particle and the state. If an electron is in the p orbital, then the electron *is* the p orbital. Quantum theory only tells us how energy is distributed into parts, and there are many ways to construe this distribution. It is impossible therefore to describe particles and objects in quantum theory without reference to the whole system; each part can only be defined via the 'role' it plays as part of a complete ensemble.

When parts are independent of the other parts, then we can reduce the whole into the parts and treat the whole as an emergent property of the parts. When the parts are only defined in relation to the whole, then the whole is in some sense logically prior to the parts. Current physical theories speak about structures as part-part relationships due to the belief that the whole is nothing but the aggregation of independent parts. The notion therefore that a part gets it properties due to its relation to the whole needs a conceptual shift in defining the whole independent of the constituent parts.

The reductionist dogma in science draws upon a classical picture of nature in which we suppose that all objects exist independent of the other objects. These objects, we further suppose, randomly combine to create structures. For instance, we suppose that a table, chair, or bed is made out of planks of wood which exist within the table, chair, or bed in the same way that they exist outside these structures. A given plank of wood could therefore be part of different collections or structures which constitute a table, chair, or bed, although this participation in different collections does not change the participating plank. If the table, chair, or bed is only an outcome of aggregating the individual planks of wood, then the planks of wood are the same whether they belong to a table, chair, or a bed.

This classical physical view of nature has undergone a radical denial in modern physical theories although the new ideology that fits the nature of reality hasn't yet emerged. Without a new theoretical grounding in our thinking about nature, other areas of science—such as chemistry, biochemistry, and biology—continue to use the classical dogma about nature. If the problems of modern physics were well-understood, their solution would require us to think of the parts as being co-dependent rather than independent. The leap in science is, therefore, that the planks of wood are not the same in the table, chair, or bed. Rather, there is something unique in each plank of wood depending on whether it belongs to a chair, a table, or a bed. Now it is not sufficient to describe the properties of the plank independent of the other planks, because each plank acquires some additional properties within a specific type of collection. The only way these properties can be explained is if we can speak about the plank as part of a bed, a table, or a chair. That in turn requires us to speak about the table, chair, or bed even prior to the planks.

The classical physical belief that the whole is nothing but the collection of independent parts leads to a problem of recursion when the parts have to be defined in a new way as portions of some whole. The problem of recursion arises because the whole depends on the parts and the parts depend on the whole. To speak about the whole we must first collect all the parts, although the parts themselves cannot be completely defined unless the whole has already been defined. The definition of parts therefore depends on the definition of the whole, which in turn depends on the definition of the parts. There cannot be a clear definition of parts or wholes due to this recursion. And, therefore, there cannot be a notion of a part that acquires a unique property as part of a collection. All the parts must have properties that are independent of the other objects in the collection and the collection is simply a random aggregation of parts.

This problem is so fundamental in current science that it forms the basis of almost every debate about structure and function in biology. The debate arises because when we look at biological forms, we see a tight interrelation between the parts—quite like how planks of wood form a table, chair, or bed. This tight interrelation prompts us to the idea that there is something unique about a part within the collection; for instance, a molecule has some additional properties within a biological being rather than in the test tube. And yet, there is no theoretical basis on which this idea can be justified. The reductionist supposes that all the parts in a whole are independent and the functional behavior of the whole must therefore be a random occurrence of the aggregation of the independent parts.

Some design theorists now argue that the aggregation of parts into a whole is not random but the result of some intelligent activity. However, this intelligent activity does not seem to have a natural cause: its effect can be observed, but it cannot be explained. A non-natural explanation of biological complexity takes us out of science, as far as explanations and predictions of facts are concerned. For instance, which forms should be created by design and why? Of course, current evolutionary theory also does not offer any predictive usefulness, so evolution cannot counter design arguments for this flaw. However, in so far as both evolution and design fail to predict diversity, they remain unsatisfactory scientific ideas.

There is only one way to solve this problem, which is to postulate that the whole exists in some sense independent and prior to the parts. This idea seems bizarre in the context of current objective thinking, but it can be made rigorous in a new semantic view.

I will have many opportunities to discuss the basis and implications of this idea throughout the course of this book; I will show how many physical theories—including quantum theory, statistical mechanics, and general relativity—show that the only consistent way to think about reality is to acknowledge that a *system* is something logically prior to the *objects* in it. However, since this idea is so antithetical to the picture of nature developed in Newton's times, it has generally been ignored and neglected within science. Nevertheless, to the extent that the dominant dogma in science (that objects are logically prior to systems) is incompatible with the implications of empirically proven physical theories (as systems being logically prior to objects), this leads to many interpretive difficulties in understanding the newer physical theories. These difficulties include the measurement problem in quantum theory, the irreversibility in statistical mechanics and the indeterminism in general relativity.

A deep appreciation of problems in physics is necessary to understand why the idea of a whole as something logically prior to parts is required to solve the problems of measurement, irreversibility, and indeterminism within physics. Similar necessities arise from the need to solve logical paradoxes in mathematics and computing, as we shall see during the course of later chapters. The notion that the whole system is logically prior to the parts in it is therefore essential to the evolution of science and I will elaborate on the reasons for this shift in thinking in subsequent chapters. However, before I dive into discussing those problems, it helps to sketch some intuitive examples of this basic idea. To see how the whole can be logically prior to the parts, consider the example of how an ordinary chair is viewed semantically. Figure 1 illustrates the development of the idea of a chair into more complex ideas.

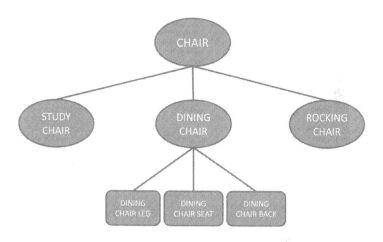

Figure-1 Refinements of Everyday Concepts

In everyday language, notions about 'study chair,' 'dining chair,' and 'rocking chair' are refinements of the idea of a 'chair.' We must first have the notion of a 'chair' before we can speak about a 'rocking chair.' As an idea, the notion of a 'chair' is more primitive and simpler relative to the idea of a 'rocking chair.' In the same way, the parts of a 'dining chair' such as 'leg,' 'seat,' and 'back' are more complex notions relative to the notion of a 'dining chair.' Unlike the 'chair' which needs just one word, and a 'dining chair' which needs two words, a 'dining chair leg' needs three words. To describe further smaller parts of the 'dining chair leg,' we would need to use even more words, such as 'dining chair leg base' or 'dining chair leg top.' As we get to smaller and smaller parts, we require more and more words to describe these parts. Note how this approach begins in the idea of a chair and then constructs various types of chairs and then their individual parts. This

approach to describing the parts leads us to the following general conclusion: the whole chair is semantically simpler relative to the parts
that make it up. In so far as we think that the simpler things are logically prior, we could now suppose that the chair is logically prior to its
legs or seat.

	PHYSICAL COMPLEXITY	SEMANTIC COMPLEXITY
SMALL	LESS	MORE
BIG	MORE	LESS

Figure-2 Semantic vs. Physical Complexity

The semantic notion of complexity overturns our approach to
reduction in nature. In the physical notion of complexity, the smallest
parts of nature are more primitive because we believe that small is
simple. Physical reductionism therefore begins with the smallest parts
and tries to build larger things by aggregation. The aggregation gives
the impression that the whole is reducible to parts, although physical
theories are unable to carry out this reduction, without dropping classically entrenched ideas such as objectivity, reversibility, and determinism. Could we now adopt a different type of reduction in which the
whole is simpler and the parts are more complex, because the parts
are created by adding more information to the whole? In the above
example of a chair for instance, the 'chair leg' is a more complex concept relative to the concept of a 'chair,' even though the leg is smaller
relative to the entire chair itself.

The clear advantage of semantic reduction is that the notion of a
'chair' exists within the notion of a 'chair leg,' although we construct
the part from the whole rather than the whole from the parts. Since
the 'chair leg' is produced from the 'chair,' the 'chair leg' is different
from the 'table leg' (which is produced from the 'table'). To formulate this difference, the parts refer to the whole, but the whole does
not refer back to the parts, because the whole is logically prior to
the parts. This relieves us of the problem of recursion in the physical
theory and of the issue in understanding how a part acquires some
additional properties within a whole. Now, the 'leg' within the 'chair'
and 'table' are not the same object. Rather, they have some unique

contextual properties which arise from the fact that they belong to different collections or wholes. The illustration shows that the problem of recursion arises only when the whole is created from the parts. Science can still be reductionist, although reductionism is itself defined in a semantic rather than a physical manner.

Essentially, in the semantic approach, the simplest things in nature are elementary meanings. A 'chair' is simpler relative to 'chair leg' because the 'chair leg' has more semantic information in it. This helps us understand why a collection—e.g., a chair—cannot be reduced to independent planks of wood because each plank of wood in itself is defined as 'chair leg', 'chair back', and 'chair seat' which in turn carry the notion of a 'chair' within them. If we think that 'chair' is the whole, then the whole is somehow present in each part. Likewise, the whole must be logically prior to the constituent parts.

Semantic descriptions wreak havoc to many current scientific ideas, but all these changes can be summarized in a single new idea—a shift in our understanding of the nature of space and time. Space and time are methods by which we distinguish objects; an object at a different location in space or a different instance in time is a different object. If differences in objects are themselves created by dividing an elementary concept by more concepts, thereby creating a hierarchy of concepts, then space and time also have to be described hierarchically rather than linearly. Common examples of such descriptions already exist in the everyday world. For instance, locations in the everyday world are described through postal addresses, which require a hierarchical nesting of countries inside the earth planet, the nesting of states inside the country, the nesting of cities inside a state, the nesting of localities inside a city, the nesting of streets inside a locality, and the nesting of houses or offices inside the street. Time is similarly described hierarchically in the everyday world by embedding years inside centuries, months inside years, weeks inside months, days inside weeks, hours inside days, minutes inside hours, seconds inside minutes, and so forth. The hierarchical notions of space and time are semantic while the physical notions are not; the physical notions about space and time are linear.

In the physical notion of space, the universal space is created as the collection of all possible locations and from these locations objects emerge. In the semantic notion of space, the individual locations in the universe are created by adding more information to the origin of universe. In the physical notion of matter, similarly, the atoms are simple things and the universe is a complex thing. In the semantic notion, the universe is simple and the atoms are complex. In the physical notion of space, we speak about structures in terms of part-part relationships and in the semantic notion of space we must speak about structures in terms of whole-part relationships. In the physical notion of space, the whole reduces to the independent parts (which are logically prior to the whole). In the semantic notion of space, the parts are created from the whole by dividing simple ideas through the addition of increasing amounts of information.

Essentially, in the semantic view, the universe is the simplest idea which is then refined through successive steps that add information to the primitives to create ever more detailed objects. The collection of all objects in the universe is the collection of all the ways in which the idea of the universe can be refined. The universe is therefore not merely a collection of things, but all the things in the universe themselves carry the notion of universe within them.

This idea can also be formally stated as follows: The universe is an inverted semantic tree whose root is an elementary meme. Refinements of this meme create myriad objects in the universe.

When space and time are constructed as hierarchies, then space is closed and time is cyclic. Postal addresses and clock times are therefore not merely artifacts of our human description of nature. They can also be reflections of some real properties in nature which cannot be captured through current physical theories because in these theories space and time are linear, open, and infinite. The semantic view changes how we formulate fundamental theories about nature because it changes the nature of space-time itself.

There are many ways to divide the universe into hierarchical trees thereby creating differing descriptions of nature. If space and time are themselves defined through such hierarchies, each hierarchy represents a unique coordinate reference frame. Each such reference frame denotes different ways in which individual locations and instances are constructed through a semantic division. Each such location must be given a hierarchical name, similar to postal addresses and clock times. Unlike the relativistic notions of space-time in current physics, which allow infinitely many equivalent descriptions of each location and instance in space-time, a hierarchical space-time must be absolute. The absoluteness of space-time implies that only one of the methods of constructing the universe is real, and can be true. For instance, whether two objects are part of the same collection or different collections must be an empirical question. Likewise, whether two instances in time are part of the same time cycle or different time cycles is an empirical question.

In current physical theories, the universe can be arbitrarily divided into collections or systems and the manner of drawing such boundaries in space-time is only a practical convenience by which we choose to study limited parts of nature. Such boundaries have no theoretical or empirical consequence to the predictions of a theory. If, however, space-time is hierarchical, then the manner in which the universe is divided into parts is a scientific question. Whether an object belongs to a structure or not is now a scientific question. Structures will now not be unique to physics or biology; rather, they will appear in every field of science, and change the way we think about objects, not as independent things but as parts of collections.

The key biological implication of these ideas is that the living world must also be described in terms of hierarchies of ecosystems and its evolution in terms of time cycles. The identity of a biological species cannot now be defined without reference to the ecosystem it partakes in. A species can only be defined in relation to the role it plays in the ecosystem. Indeed, as I will discuss shortly, no single species or individual in that species can change without a larger ripple effect within the entire ecosystem. In fact, all individual changes in the ecosystem will be reversed unless the ecosystem as a whole evolves. The evolution of the ecosystem as a whole and its individual parts must be described cyclically rather than linearly.

The difference between the current evolutionary theory and the alternative that I will discuss in this book is therefore based on the following two distinctions: (1) parts vs. wholes, and (2) linear vs. cyclical. Current evolutionary theory describes the evolution of species as a linear evolution of parts. The alternative that I will propose is a theory that describes the evolution of wholes in cycles. These two changes follow from a revision of our notions about space-time from being 'open' and 'linear' to being 'closed' and 'hierarchical.' These changes in turn follow from the problems of indeterminism, measurement, and irreversibility in physics, and the problems of incompleteness and incomputability in mathematics and computing theory, respectively. These problems are in turn closely related to the issue of structure and meaning in biology and to the problem of meaning in the mind and its expression in natural languages, as we shall later see. The current theory of evolution is based on a linear and open notion of space-time while the alternative that I will describe in this book is a semantic theory of evolution that involves closed and hierarchical notions of space-time.

The semantic view solves many problems in science but it also overturns the dominant dogma about objectivity. Now nature does not begin in objects; it rather begins in ideas which exist in nature as systems. The current theory of evolution inherits the same kinds of problems of incompleteness, uncertainty, indeterminism, and incomputability as the other physical and mathematical theories built from the idea of objects as the basic building blocks of nature. A deep understanding of the problems in physics, mathematics and computing helps us see how these problems also affect biology and the theory of evolution. It also helps us see why the semantic view of nature— that solves the problems in physics, mathematics, and computing— also fixes them in the context of biology and evolution.

The Shift in Science

Many of the arguments in this book are developments of my prior work in mathematics, computing, and quantum theory, which are elaborated in my earlier books *Gödel's Mistake: The Role of Meaning in Mathematics* and *Quantum Meaning: A Semantic Interpretation of Quantum Theory*. This book summarizes those arguments and extends them in the context of biology and evolution. My earlier work discussed the reason why current mathematics is incomplete: numbers are properties of collections and not properties of individual objects, although mathematics tries to construct numbers from objects. This leads to the problem of recursion described above because we cannot speak about an object being "the 5ᵗʰ" inside a collection without referring to the collection itself. A consistent number theory requires the collection (a set in mathematics) to be defined independently of the objects that make it up, but we cannot do so, if we think that collections are created by aggregating objects.

To solve this problem, collections must be treated as concepts, such that they can be defined without reference to member objects. My earlier work also shows how the quantum indeterminism problems are solved if quantum objects are defined in relation to the collections: if we could describe the collection prior to its members, then it would also be possible to speak about the order of objects within that collection, alleviating the problem of probabilities which arises because we cannot predict the order of quantum events. Finally, the previous books describe why the problem of meaning in computing cannot be solved in the current theory of computing because the meaning of a program instruction is only given in relation to the program as a whole. Computers that look at the individual instructions cannot know if the program is useful or malicious.

The problem of the whole-part relation is old, deep, and widespread. It underlies the issues of meaning in linguistics, the problem of mind-body interaction in neuroscience, and the debate around universals in philosophy. I briefly outlined above the incarnations of this problem in the context of mathematics, computing, and physics. This book is specifically devoted to this problem in the context of biology. The realization that the problem is old, deep, and widespread is necessary to understand why there cannot be individual, separate solutions to this problem. Rather, the scheme that solves the problem in one situation must be used to solve it in others. In that context, the book discusses how this problem appears in other areas to identify the common issues and the nature of the solution.

The thesis of this book is rather bold. The thesis is that there are additional properties in nature which arise when a scientific theory attempts to describe collections rather than individual objects. These properties cannot be reduced to the objects in those collections. Examples of these properties include the notion of numbers, which cannot be defined without a collection, and the meaning of a computer program which cannot be understood without taking all the program instructions into account. Such properties also underlie the problems in understanding physical systems when theories describing these systems attempt to reduce nature to their constituent parts. The attempt to reduce the properties of a collection to the properties of the independent parts in that collection results in incompleteness, indeterminism, and probabilities. A new way of looking at collections and parts is now needed. This new way, I will argue, requires us to construct parts from wholes rather than wholes from parts. Biology is part of this shift in thinking; in the context of biology, the problem of collections advents as the problem of structures and functions. The solution is that if atoms themselves are defined only in relation to the collection, then structures and functions too are not epiphenomena of independent parts. Rather, we must now define structures and functions even prior to defining the objects that make up the structures and functions.

This shift in science can arise from a redefinition of space-time as being hierarchical and closed rather than flat and open. Objects in this space-time have contextual properties because each location in space-time is defined only in relation to the next hierarchical location and instant. Like a 'rocking chair' is produced from a 'chair,' and a 'rocking chair leg' is produced from the 'rocking chair,' similarly, parts of the universe would be produced from the universe. It is impossible to conceive of the whole as being logically prior to the parts in the physical view of nature. However, it is possible to think in this way if the universe is ideas. As ideas can be refined and elaborated to produce more complex ideas, similarly, if things were ideas then they could also be produced by refining other simpler things.

The shift in thinking requires us to construct objects from meanings rather than meanings from objects. It should be noted that a theory of meanings is also a theory of objects, but not vice versa. This is because a physical representation of meanings—a symbol—is also an object, although objects are not necessarily symbols. A physical theory of objects defines objects as independent entities while a semantic theory of symbols defines symbols as part of collections. Accordingly, a theory of the evolution of symbols is also a theory of the evolution of objects, but a theory of the evolution of objects is not necessarily a theory of the evolution of symbols. The semantic view will therefore not deny mutations or selection. However, it will deny the idea that mutations are random and that the selection is physical (i.e. based on physical properties). The alternative will show that mutations and selection are based on semantic principles, which in the context of biology can be equated with structures and functions. The semantic properties of structures and functions cannot be studied in the current physical theories, but they can be understood by using a new approach to space-time.

The intent of this book is to describe the physical, computational, and mathematical issues that arise from the existence of meanings, connect these issues to structure and function in biology, and generalize this idea into a different view of space and time.

I will show that the issues that the current theory of evolution is trying to solve are fundamentally unsolvable in the present physical view of nature. The solution requires a new approach to studying matter, in which objects are symbols, whose meaning is given in relation to the wholes, whose evolution is cyclic. These ideas will then be connected to the problem of biological evolution. This line or argument, I believe, is quite original and therefore a useful addition to the ongoing debates on the origin of species. The second chapter will elaborate on the objections to evolution from physics, mathematics, computing, and game theory. The third chapter will discuss an alternative that I believe holds the promise of making empirically testable predictions in areas outside biology. The fourth chapter describes the applications of this alternative to biology and how it entails a newer model of evolution. Since this alternative is yet another viewpoint in a growing list of alternatives to evolution, the fifth chapter will contrast the new view previously described with the ideas that have already been put forward by others.

However, before I get started with a detailed analysis of these issues and their proposed solution, I want to develop some basic intuitions that motivate the whole-and-cyclical view of evolution.

The part-and-linear view of change in current science is a legacy of classical physics which describes the motion of independent particles. It is generally believed that classical physics suffices to describe the everyday world, although classical physics would evidently not describe the biological world. The assumption therefore that the everyday world is a classical world is clearly false in the case of biology. The reason for the failure of classical physics in biology is quite profound: the everyday world is not a classical world. This becomes even more evident as we look at commonplace examples of semantics in the everyday world, which cannot be captured by classical physics. The evolution of complex semantic systems follows the whole-and-cyclic rather than part-and-linear model in classical physics. A quick glance at some of these illustrations will help us formulate everyday intuitions about a new model of scientific description. These intuitions are by no means rigorous or complete. However, to the extent that thinking about billiard balls and water waves helped formulate ideas about particles and fields in physics, everyday examples of whole-and-cyclic dynamical phenomena can help as springboards to a formal theory of this kind.

Linear and Non-Linear Dynamics

The idea of closed and cyclic systems is often encountered in non-linear systems which propagate a cause's effect back into the cause after a delay. To propagate the effect of a cause back into the cause, the system in which these causes and effects occur must be closed. This mechanism is generally called 'feedback,' and there are two kinds of feedbacks—positive and negative—depending on whether the effect to a cause will further reinforce the cause or suppress it. In case of positive feedback, the system very quickly spirals out of control, eventually destroying itself. In case of negative feedback, the system quickly returns back to its original state, nullifying the cause. If this model of change is applied to evolution, there are two possibilities. First, a random mutation causes a positive feedback loop thereby destroying the entire ecosystem. Second, such a mutation results in negative feedback nullifying the mutation itself. In either case, there cannot be a progressive evolution of species.

Evolutionary theory avoids the problems of non-linearity by supposing that large changes emerge through smaller changes, and these changes gradually accumulate over time. This view is based upon a one-way notion of causality where a change results in a cascading effect like a set of falling dominos. In one-way causality, the effect does not act back on the cause, and this causal model is therefore also called *linear*. A linear system is also *open*. In fact, the effects of non-linearity can only be observed within a closed system where parts of an inter-related system of parts seem to act mutually. The inability to clearly separate cause and effect is an indication of a deeper problem that the system is in itself not truly separable.

The evolutionary model of slow and gradual change therefore hinges upon an assumption that nature is open and linear. Fundamental physical theories too are linear—they build the whole system by adding up the independent parts—and these theories therefore run into deep problems when the whole is not reducible to the collection of independent parts. There is hence a deep connection between the evolutionary idea of slow and gradual change, the open and linear model of causality in physical theories, and how this model fails when applied to physical collections. The reason for failure is rather profound—when we study systems and wholes, there is something unique about the wholeness which cannot be reduced to the parts, *if* the parts were defined to be independent things.

Non-linear systems are extremely sensitive to initial conditions. A small amount of change in the initial condition can dramatically change the outcome, even when the mathematics governing the system is completely deterministic. In other words, a small change in the initial conditions does not imply a small change in the outcome. This property of non-linear systems is called *deterministic chaos* or simply *chaos*. Furthermore, a small change in one part of the system can dramatically affect all other parts. The evolutionary notion that small changes accumulate over a long period of time to become big changes is not true for complex systems because in such systems small changes do not necessarily cause a small outcome. Non-linear systems exhibit a feature sometimes called the *Butterfly Effect* due to which a small change in one part of the world can cause a big change in the other parts. Linear causes and effects have additive properties but non-linear causes and effects do not.

In a non-linear closed system, effects travel outwards from the origin of the cause and are reflected back to the origin after some delay. When the effect is reflected back, it can amplify the cause or dampen it. If the cause is amplified, the system will bounce in unpredictable ways; in the case of evolution this bouncing will destroy the ecosystem. If the cause is dampened, the system will remain stable; in the case of evolution, the dampening will reverse or eliminate the mutant. In either case, we do not expect new species to be formed by this process. Every system has in-built feedback loops in which a cause returns back as an effect. Generally, in stable systems, feedback to any change is always negative, as the system is already accustomed to certain types of behaviors. The larger the system, the more stable it is, and the harder it is to change. The larger systems do not evolve through random, incremental mutations. They can be changed only through disruptions, paradigm shifts, and revolutions in which large, if not most, parts of the system are changed at once.

Once a large system has been realigned, it takes a while before things settle down to a new stable point. In complex systems, there are always points of stability such that any perturbation from the point of stability will be quickly reversed, returning the system back to its original point of stability. The only other option in complex systems is for the system as a whole to leap from one point of stability to another, and these points—in a large system—may not be close by. That is, if the system as a whole has to leap from one stability point to another, then it must make a large concerted change.

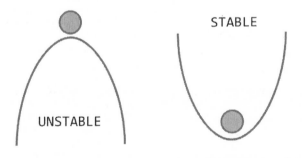

Figure-3 Stability Points in a System

The problematic feature of complex systems is that if such as system has moved sufficiently out of its stability zone, such that it cannot revert back to its stable state, then it will be chaotic. Around a stable point, it is hard to take the system out of stability. But if the system has been taken out of stability then finding the next stable point is not trivial. There is nothing that will easily drive the system from one stable point to another. But if such a change were to occur, it would require a large, concerted change in the whole system.

Perturbations can be used to understand the *stability* of a dynamical system. For instance, a ball lying at the bottom of a pit can be pushed to the left or right but it will always settle at the bottom. If the ball is at the top of a bell-shaped surface, then it will fall down into a stable state when perturbed. The bottom of the bell curve is a point of stability and perturbations around the stability point will send a system back to its stability point. Perturbations create temporary ripples in a system but the ripples die down when the system returns to normal operation. The points of stability—which 'attract' the entire system into a stable state—are also called *attractors* in the theory of complex dynamical systems. Attractors are subsets of the entire phase space of a dynamical system; a phase space represents the space of all possible states a dynamical system can possibly occupy. The existence of an attractor implies that not all possible states are equally likely. Rather, some states are much more likely than others. The system as a whole enters these states, and remains there. Small perturbations may momentarily take the system out of stability, only to quickly return it back to the same state.

Perturbations, however, do not explain how a system goes from one stable state to another. A large number of changes must happen at once to cause a systemic shift from one attractor to another. A single random mutant cannot cause a systemic change; rather the mutant will be wiped out by that change. Thus, large-scale changes cannot emerge via incremental random mutations. In an ecosystem, a number of diverse species are interlocked in exchanges of give and take, which creates a system *resilient* to change. Individual mutations in the ecosystem will not work, even for the mutant, because the outcome of the mutation is incompatible with the rest of the system. If incompatible mutations occur, the environment will eliminate the mutant. No mutation in an animal species gives the species an advantage large enough to change the entire ecosystem into a new state. The natural conclusion is that any individual mutation will only adversely affect that mutant. The only way a mutation is viable is if the entire ecosystem benefits from it, which requires a large number of simultaneously changes to occur.

The idea of random mutations leading to gradual and incremental evolution flies in the face of the problem of chaos, because random mutations are expected to deliver locally significant changes over geologically significant time periods, while chaos tells us that globally significant changes can be delivered in very short time periods. The idea, therefore, that if the giraffe has a longer neck it only affects the leaves high up in the trees is false. It is more accurate to suppose that if a giraffe has a longer neck then it will affect everything in the ecosystem in a very short period of time, in ways that we don't yet know how to predict. This effect has three possible consequences: (1) it kills the mutant giraffe, (2) it kills the entire ecosystem, or (3) it evolves the whole ecosystem to a new state.

The theory of evolution takes a peculiar approach to the relation between the living being and the environment. Rather than look at what happens to an individual when it mutates in a stable environment, it looks at what happens to the individual when the environment has already changed. If we assume that the environment has changed, then the change will push the individual in that environment to adapt. However, a closer look reveals that the environment is as much comprised of individual mutations as the living beings. Therefore, there is no grand environment that will change unless the individuals in that environment change first. When some individual changes in a stable environment, the most common result is the elimination of that mutant. Therefore, the only way the environment as a whole can change is if lots of individuals in that environment change simultaneously. That is clearly not randomness.

Evolutionary theory rests on a less than proportionate response to a change. In this case, the environment changes, and the biological species adapts to it and those who cannot adapt would be eliminated by the environment. This is a one-way view of causality in which the environment acts but the living being does not react back to that change. If the environment and the living being are equally likely to mutate[1]—because changes in both are equally governed by random mutations—then the system will either oscillate violently or the original mutant would be quickly eliminated.

All modern physical theories are linear, which is an outcome of the assumption that the system is *open* and *deterministic*. Evolutionary theory changes one of these assumptions—namely that of determinism—to postulate random mutations without considering the other assumption of openness. When randomness is combined with openness, it leads to the idea that small changes will eventually lead to big changes. However, if we discard the assumption of openness and consider closed systems, both deterministic and random mutations lead to oscillations. As the amplitude of these oscillations increases, the entire ecosystem will be destroyed, unable to handle the fluctuations. If the amplitude subsides then the ecosystem enters a stable state, unable to create new species. The only alternative is that the ecosystem as a whole shifts from one state to another, which, however, cannot be explained by random mutations.

The key theoretical question that arises from this situation is: Should we describe a biological system as open or closed? An open system does not have feedback loops, while a closed system does. An open system is described by linear equations, while a closed system is described by non-linear equations. These two types of systems have some dramatic differences. While a linear system evolves gradually and incrementally, a non-linear system changes radically and may oscillate violently. The theory of evolution uses natural selection for feedback but does not change the theoretical model from linear to non-linear dynamical systems. The theory supposes that there is gradual evolution resulting in large changes over time based on a linear model, rather than taking into account the consequences of non-linear systems, which oscillate violently in case of positive feedback and prevent change in case of negative feedback.

I will later discuss a solution to the problem of linear and non-linear theories in which the whole is a linear sum of parts but the parts are not *independent*. Rather, the parts are defined in relation to the whole. This approach allows us to have a linear theory of the whole, while giving us the ability to explain non-linear phenomena. The whole is a linear combination of parts, but we construct the parts from the whole rather than the whole from the parts. As noted above, the shift in the whole-part relation can be understood in a semantic view where the small is created from the big by dividing the big into parts through the addition of information. Now it is possible to construct a predictive theory of evolution, although this theory will predict the evolution of parts as a *consequence* of the evolution of the whole rather than vice versa. In other words, it is not individual mutation that leads to large-scale changes. Rather, it is the large-scale changes that result in the individual mutations.

Evolutionary theory can be revised in light of these ideas. The current view is based on the idea that each part of the system is independent of the other parts and therefore mutates individually. A closer look however tells us that if the parts of a system were to mutate independently then the mutations would almost always be reversed due to a non-linear feedback loop, or the system would oscillate perennially destroying the ecosystem in the process. The only way a biological ecosystem can evolve without destroying itself is if there is a linear theory of the evolution of a system as a whole. In this theory, the parts are constructed from the whole, and when the whole evolves, the parts also evolve consistently. To describe the evolution of all independent parts we must postulate a scenario in which all the parts co-mutate to create consistency, which is highly unlikely. However, if we describe the evolution of the whole (that constructs the dependent parts) the evolution can be linear, deterministic, directed, and therefore predictable. My interest in the ecosystem view stems from the dilemma between linear and non-linear approaches. Specifically, I wish to show that if we think of the whole ecosystem rather than its parts, there are some notable conclusions that can be drawn, which are not evident from a theory of parts.

Ecosystems and Ecology

It helps to illustrate the above problem arising from closed and non-linear systems with an intuitive example, and I will presently try to do that. An ecosystem can be modeled as a cycle comprising a large number of species and the environment comprising soil, water, air, and sunlight. In this cycle, one living being is the food of another. At the bottom of this food chain are plants which subsist on water, minerals, and sunlight. These plants become food for herbivores, which in turn become food for carnivores, which are food for scavengers and detritivores, which convert biological material back into the soil, water, and air that in turn fuel the growth of the plants.

An ecosystem that combines many such species in a cycle of nutrient circulation has to run at a specific rate for the living beings in the cycle to survive. If the cycle runs too fast or too slow it will kill most living beings because these are accustomed to inputting and outputting matter at a certain rate. For instance, if there is too much water or too much sunlight (or too little water or too little sunlight), all species downstream from that event will be severely affected. This input-output rate is a basic property of the ecosystem, and any changes to it will affect the entire ecosystem, requiring all species in it to change all at once. It is not possible to change a single property in the ecosystem without changing its stability point as a whole.

The key point of considering ecosystems is therefore that species are closely interlocked into input and output exchanges. Any change to the input-output relation in any part of the ecosystem does not remain limited to that part alone. Rather, changes in input-output relations in one part of the ecosystem ripple through the entire ecosystem due to the interconnected nature of the ecosystem.

For instance, let us suppose that a carnivore species undergoes a mutation because of which its appetite is increased. By this mutation the carnivores eat the herbivores faster relative to the current rate of herbivore replication. If the herbivores don't replicate much faster, their population would soon be reduced. This in turn will affect the plants upstream from herbivores and the carnivores downstream. The plants that were being eaten by the herbivores will now replicate much faster (since the herbivores that ate them earlier are significantly reduced) putting far greater demands on the supply of minerals and water. Similarly, the carnivores will not have enough herbivores to subsist on. Decline in herbivores will cause an increase in plant population but a decline in carnivore population. These two changes will in turn ripple throughout the ecosystem in new ways. Since the herbivores are a link in the ecosystem cycle, changes to their population will affect all aspects of the cycle.

If the herbivore species is to avoid extinction, it must start to replicate faster. This would in turn imply more sexual activity, more food, faster biological development, etc. on the part of herbivores. Since the herbivores subsist on plants, the increased biological activity in the herbivores will also require the plants to accelerate their biological activities. But faster plant biology also requires faster soil regeneration and a greater water availability, which in turn depends on a faster geological cycle. The only way a greater appetite in the carnivore will keep the ecosystem intact is if the entire ecosystem runs at a faster rate. That implies a change in most if not all parts of the ecosystem. In a tightly intertwined cycle, any change in the cycle ripples through and impacts every living being in it.

Of course, when an entire ecosystem has to evolve simultaneously, it is possible that at least some parts of the ecosystem will not change. Some dramatic consequences follow when this happens.

For instance, if the carnivores develop a larger appetite but nothing else changes in the ecosystem, the carnivores will quickly run out of herbivores. Each carnivore will have a larger appetite but a lesser overall amount of food. Now, the carnivores themselves have to struggle to find food, and those who can live with lesser food will survive better than those who need more food. The carnivores which underwent mutation are now less suited to survive in the modified environment as compared to those which did not, because the carnivores requiring more food have a greater amount of struggle as compared to those carnivores which could survive on lesser food. Thus, if an animal undergoes mutation but the environment essentially stays constant, then the mutant is disadvantaged in this stable ecosystem. These mutants would now be eliminated by their own change, if principles of natural selection were applied to them. For instance, the animal species which developed a larger appetite would be less suited to survive relative to those that did not.

Of course, we can suppose that a reduction in the food quantity is not a problem for a species if some members of the species evolve further skills to outmaneuver the others with the normal appetite. For instance, the animal with a larger appetite could become faster, stronger, and bigger to fend off its competitors. But now there is a direct competition between the survival of those animals which have a lesser appetite versus those who are becoming faster, stronger, and bigger to support their larger appetite. Simply having a larger appetite isn't therefore enough; these changes must be accompanied by other changes in the same animal that make the acquisition of food more likely than before. Again, many changes must happen simultaneously even for the mutating animal to survive.

Suppose that there are faster, stronger, and bigger animals which have grown due to their bigger appetite, and they can manage to outmaneuver the others for food. The animals which aren't stronger, faster, and bigger would be eliminated in this struggle and the offspring of the stronger, faster, and bigger mutants will continue to live. Since the total amount of food is constant, the total population of the mutant species with a larger appetite must be smaller as compared to the population of the original species with a lesser appetite. Increase in appetite therefore leads to a decline in the carnivore population, all other things being equal. When the carnivore population has decreased, its density per geographical area reduces. The species is now more susceptible to attack from other species, or other calamities, and will need methods to cope with the risks.

To avoid destruction, the mutant species must become even bigger, faster, and stronger to fend off potential predators, which means that they now need even more food. This additional demand for food, which arose from the problem of the reducing population, will now create an even greater competition amongst the remaining members of the species, causing a further decline in their population. This in turn will make the population even more susceptible to predators, which in turn will mandate the remaining members to become even faster, stronger, and bigger, which in turn requires more food, which creates more competition, which leads to an even smaller population, and the cycle continues indefinitely. In the end, the mutant species will be wiped out as a consequence of its own growing appetite and the subsequent increase in competition.

When the mutant carnivores have been eliminated, the herbivores begin to thrive, as there is no one left to eat them. These herbivores have an abundant supply of plants left over from the time when there weren't enough herbivores. Soon, the herbivore population increases and the plant population decreases. The elimination of the carnivores therefore leads to a dramatic increase in herbivore population, and a dramatic decrease in the plant population. To restore the ecosystem balance, there is now a great need for carnivores. If there were inverse mutations in the carnivores, such that the carnivores can now subsist on lesser food, the system would return to its original balance between carnivores, herbivores and plants, after undergoing an oscillation as is shown in Figure 4.

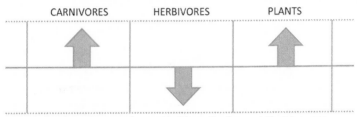

A: Carnivore Increase Causes Herbivore Decline and Plant Increase

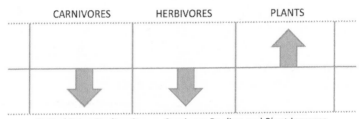

B: Herbivore Decline Causes Carnivore Decline and Plant Increase

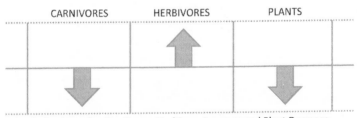

C: Carnivore Decline Causes Herbivore Increase and Plant Decrease

Figure-4 Ecosystem Population Oscillation

When the carnivore population booms, the herbivore population goes bust and the plant population booms. When the carnivore population goes bust, the herbivore population booms and the plant population goes bust. Also, a boom is followed by a bust, because, beyond a certain point of growth, the boom is unsustainable. These cycles of boom-and-bust are not isolated events in a particular species but must ripple throughout the entire ecosystem, causing other peaks and troughs. The system evolves as a whole, and it evolves cyclically. Of course, it is not necessary that such a boom-and-bust cycle should occur. As I noted above, a carnivore species with a lower appetite is better suited in a stable environment. That alternative, however, implies that the mutation of a larger appetite will almost always be reversed, since the ecosystem remains constant, unless the ecosystem as a whole itself undergoes a transformation.

There are hence two alternatives in an ecosystem. First, small mutations would be immediately reversed as a single mutation is not capable of causing a large scale change; e.g., a simple increase in appetite would not work, unless there are also concomitant changes that make the animal bigger, faster, and stronger. Second, if the individual mutations are not reversed, it will result in ecosystem oscillations; e.g., if the animals become bigger, faster, and stronger, and acquire large appetites, then the population will oscillate. In this oscillation, the mutant species would be eliminated causing a rapid rise in the population of other species which the mutants previously consumed, which in turn causes a rapid decline in the population of other species downstream from the consuming species. As the latter population declines, the consumers begin to decline, too. Eventually, the ecosystem returns back to normal only when the mutants have been completely eliminated, and the cycle has been reversed.

In a *stable* ecosystem, when the environment is constant, any unilateral mutation in a species will make the mutant less likely to survive relative to the other members of the species which do not mutate. If the mutant has to survive, then it must develop successive mutations which are increasingly improbable and further decrease the chance of survival unless even more improbable mutations are applied. In other words, in a stable system, as a species moves away from the point of stability, its existence becomes increasingly unlikely. The limit to this unlikeliness is that the species will be eliminated. Of course, this would generally not happen, because there are more likely outcomes, such as the reversal of the mutation, which return the system back to its original stability point. Thus, the mutations must be reversed to return to stability, or the species will be completely eliminated. Naturally, it is better to suppose that the inverse mutations become more beneficial than the forward movement. The only way any mutation will work for a species is if most of the species mutate at once to change the ecosystem as whole.

In any ecosystem, the macro picture of evolution is that a system as a whole evolves to a new state. The micro picture of evolution is that after a macro change, the individual species in that ecosystem would adapt to the macro changes. Individual random changes will always be reversed by the ecosystem. Only collective and large-scale changes will cause the ecosystem as a whole to evolve, and cause the smaller changes in it to occur. However, the collective, simultaneous, ecosystem level change is highly improbable if these changes are seen as aggregations of individual change. A different type of theory that explains macro changes is needed.

Adaptation vs. Retaliation

Each species in the ecosystem is equally likely to undergo mutations, but no species has a better chance of that mutation surviving. If species A mutates unilaterally, thereby impacting species B, we cannot predict if B will adapt to A, or A will be eliminated because B retaliates to A's changes instead of adapting to them. For example, if giraffes have longer necks to eat the leaves shouldn't the trees become even taller to retaliate in competition with the giraffes? Thus, even if we assume that there is ongoing random DNA mutation in both giraffes and trees, it is not clear which of the competing DNA mutations will survive. Regardless of the physical differences a species may have from other species, those physical characteristics don't make DNA mutation more effective in one case than in another. Mutation depends on the DNA and the DNA in one species is not more likely to mutate relative to the DNA in another species.

If every species is equally likely to mutate for its own benefit, then there is a low chance for a mutation to be successful, because the disadvantaged species may retaliate against the advantaged one, thereby nullifying the evolutionary consequences of mutations. Indeed, as I will shortly discuss, the cancellation of these mutations are indicated by game theoretic considerations which indicate the Tit-for-Tat strategy to be the best strategy for long-term play.

The only exception to this cancellation is if there is a *set* of mutations that collectively benefit the ecosystem as a whole, thereby creating a win-win situation for most species. For instance, suppose that there is a mutation in trees that dramatically increases their number of leaves, making the branches on the trees unable to manage their own weight, accompanied by another mutation in giraffes that dramatically increases their sexual activity creating more giraffe offspring, who now need more leaves. These two mutations will mutually benefit giraffes and trees because trees are better off losing their leaves and giraffes are better off eating those leaves. In fact, this is an example of complementary pairing of problems and solutions. In this case, more leaves are a problem while more giraffes are a solution to that problem (we could inversely claim that more giraffes are a problem and more leaves are the solution.).

Biologists call this mechanism of change *coevolution* because two or more species change their traits simultaneously. It should be recognized that coevolution between a pair of species causes other effects to other species, and those effects have to be acceptable to those species as well if the change is to survive. At least, there has to be a large enough mutually beneficial synchronous change in which the benefits outstrip the disadvantages. Without a coordinated change, a single mutation will most likely cause changes which other species will retaliate against and thus eliminate the mutant.

The pairing of changes in coevolution—such as the concurrent changes in giraffes and trees—only illustrates the nature of the solution, but it is hardly a complete solution in itself. In reality, more leaves on trees requires more water and soil nutrients for the trees, which require more rains and a faster recycling of waste produced from giraffes into soil. The problem isn't therefore entirely solved without a concerted modification of trees, giraffe, soil, water, and climate. The story doesn't end here because there are factors that determine the nature of soil and the availability of water. For instance, high levels of pollution could contaminate the soil and make the water useless. The problem now widens to the nature of pollution, and how that is being caused. And even this is not the end. The problem of pollution extends into many other factors such as industrialization, volcanic activity, forest fires, and so forth. We can see that a mutation that is likely to survive is a mutation that collectively benefits most of the ecosystem. But a mutation that requires multiple simultaneous changes becomes increasingly unlikely if all these changes have to be brought about through randomness.

Evolution cannot effect unilateral changes in a single species because species are closely connected in an ecosystem. For a change to be permanent, it must ripple throughout the entire ecosystem, and there are many hurdles in that process. If all species are equally likely to mutate, and there is no direction in evolution, then two species in an ecosystem can mutate in opposite ways, forcing each other to adapt in contradictory ways. Neither species is better off in this competition in the longer run, although, in the shorter run, until the other species has had an opportunity to react to that change, it might seem that one species has a better payoff than the other.

Evolution and Economics

Of course, this is not a unique fact about biological ecosystems; the fact is seen in action in every type of ecosystem. Take, for instance, the economic ecosystem. This ecosystem connects producers, distributors, consumers, and recyclers. If any one of these players slacks, the entire ecosystem suffers. For instance, if the distribution system slows down, the prices go up, the consumers have less of the products, the producers are piling inventory but their stock value is rising. When the distribution system is normalized, there is a glut of the product in the market, the prices crash, the producer's inventory reduces but the producer's stock value also generally takes a downward hit. Given a finite size of the ecosystem, the ecosystem cannot run slower or faster, unless most parts of the ecosystem change at once. For instance, more food grain production will not work unless people start eating more, working more, and their bodies adapt to this new type of high-input and high-output lifestyle.

Even a shift in one part of the ecosystem will work only through a concerted change in many parts of the ecosystem. For instance, the only way that Apple smartphones will replace Nokia phones in the market is if the producers, consumers, distributors and recyclers shift at once. Such economic ecosystem changes are called 'market transitions' where the entire ecosystem changes simultaneously. If you are too early or too late to the transition you tend to fail and companies that act too early or act too late are eliminated by the rest of the moving market. The key point of this example is that ecosystems evolve in concert and not by incremental changes. The system as a whole hops from one stable state to another. Players who don't adapt to this change are wiped out in the transition.

Another illustration of this fact can be seen in the relation between citizens and governments through the payment of taxes. We think that we will be better off if we had more money. To have more money we could stop paying taxes to the government. That seems like a great simple change that we could all make, and which will in turn improve our chances of survival. But the environment—the government in this case—will not accept that change. The only way lower taxes can work is if there is lesser crime, lesser disease, lesser old age, lesser geo-political uncertainty, and lesser natural disasters. If all these problems were to significantly reduce then all the people currently employed in solving these problems will also become unemployed. All the businesses trying to solve healthcare, old-age, crime, defense, and natural disaster problems will now become unnecessary. And all these people now need to find alternative employment. These alternatives must be easily available for the unemployed because otherwise society will revert back into crime, corruption, war, and social unrest. My unilateral non-payment of taxes can work only if there is a huge systemic change accompanying it, which transforms the society as a whole to a safer, healthier, and more peaceful place. The larger and more interconnected the ecosystem, the more resilient it becomes to any kind of change, because there are so many more vested interests holding it back to the status quo. No unilateral change can survive in an ecosystem because the ecosystem will retaliate and eliminate the unilateral mutant, much like the government tax authority would punish the tax evaders. An ecosystem therefore has a collective destiny, and evolution must explain how the ecosystem as a whole evolves to new states.

This clearly illustrates the flaw in evolutionary theory. The problem is that even if we grant random mutation and natural selection, the theory cannot work in an ecosystem because each member of the ecosystem is equally likely to mutate and there is no direction in evolution. The different members of the ecosystem can therefore mutate in opposite directions, thereby nullifying the advantages of a single mutation. In fact, when species are locked in an ecosystem, any unilateral change will create the kinds of disturbances that will eliminate the mutant. The only way evolution can work is if the ecosystem as a whole has a *direction*—i.e. if all members coevolve for mutual benefit. But that idea would violate the fundamental premise in evolution, namely that evolution does not have a direction. Of course, I'm not suggesting that the ideas of random mutation and natural selection are themselves flawless. I'm only pointing to the fact that even if we grant random mutation and natural selection, and if there is an ecosystem, we still cannot explain evolution.

Organizational Dynamics and Evolution

Yet another well-known illustration of resilience in large systems is observed in organizations with thousands of people. All organizations develop specific types of 'cultures' in which any deviant behaviors are eliminated. The deviations may be good for the organization as a whole, but they would require the members to adapt to new ideas and behaviors, and the members will not do that. They know that any individual variation on their part will lead to discord with other established parts of the organization; that a change would result in enormous struggle and resistance with others who want to maintain the status quo. That struggle in turn requires skills and empowerment which they may not have. Rather than upset the apple cart, they would rather tag along with the majority. Large organizations thus become antiquated and rigid as they are unable to adapt to new concepts and continue to thrive on past glories.

Change in such situations arrives in one of two ways. First, it can arrive through a leadership initiative which reshapes the entire organization simultaneously. Second, it can germinate through the formation of cliques or sub-groups which nurture new ideas and practices and which are protected from the organization's dynamics. Neither of these mechanisms represents a random mutation followed by a natural selection. These rather represent the fact that when a system becomes rigid, unproductive, or archaic, internal and external forces will either break and reshape it as a whole, or if these parts are unable to reshape it they will break away to form a new organization or a new sub-culture within that organization.

In this larger narrative about large scale systemic change there are also narratives about individuals adapting to these big changes. The large scale change is the 'environment' and the individual is the entity that adapts to that environment. However, it must be noted that the individual narratives of change come into play only after the environmental level changes have occurred. Without the big systemic changes, the individual changes will be forcibly reversed because the system as such is resilient to change. On the other hand, if big systemic changes are occurring, any individual resistance to that will also be eliminated. The driver of the change is therefore not the individual mutant. The driver is the environment as a whole, while the individuals adapt to that change. This raises some fundamental questions about evolution: If the driver is the collective environment and not the individual, then how are the changes caused?

This question is the biggest one in evolution. Current evolutionary theory speaks about adaptation to the environment through random mutations. But it then attributes the changes in the environment itself to some other random events. The problem in this viewpoint is that in any large system, there must either be collective change or no change at all. A random mutation does not tell us how the collective change takes place. And a random mutation claims to aggregate into large changes over long periods of time, which is false because all such mutations—without a systemic change—will be promptly reversed or the mutant will be eliminated. This does not mean that there aren't any random mutations, but only means that they don't play the kind of causal role that evolutionary theory attributes to them. Random mutations are adaptations to large scale systemic change. They become important *after* the large scale changes have occurred, and they aid in the survival of the individual in the environment. This is, however, not the big picture of evolution. The big picture is that the environment itself undergoes dramatic changes as a whole, causing the creation of new structures. The individuals have to fit and adapt to these structural changes.

Every system brings with it particular patterns of behaviors, as it defines the roles and expectations of each part in the system. An individual change will make the mutant incompatible with the system and the change has either to be reversed or the mutant will be eliminated. The mutant may join a system which accepts its mutation, or the mutation has to wait for a system as a whole to change.

Current evolutionary theory focuses on the microscopic picture of change and attributes the creation of new species to this change. This is not a description of what really occurs in nature. The real picture of evolution is that there are massive macroscopic changes over time, and the microscopic changes adapt to the macro change. Fossil evidence also supports this because fossils show periods of massive change followed by long periods of stasis. This cannot be understood if we attribute evolution to random mutations. It can only be understood if there is a theory of massive macroscopic change within which there are small individual adaptations. If we can predict the macroscopic changes, then we can predict the creation of new species, and the direction of each individual adaptation. If we only look at the microscopic picture, and the macro system is unchanged, we can be certain that no mutation will be successful.

The dogma in current science is that the whole is built up of parts. Indeed, there isn't a whole aside from the parts. The universe is created as parts, and the random aggregations of these parts create wholes. If we stick to this dogma, then there will never be a predictively useful theory of evolution, because the facts of evolution require us to speak about the macroscopic changes as being logically prior to individual mutations. A predictively useful theory of evolution requires a shift in the current dogma of science, where we treat the wholes as more real than the parts. In current physical theories, the whole is created from the parts. There can, however, also be a theory in which the parts are created from the whole.

Once the whole-part relation has been redefined, the solution to the central problem of evolution would also be obvious: to develop a theory of the evolution of the wholes independent of the evolution of parts. In this theory, the part will evolve within the whole, which evolutionists call adaptation to the environment. However, beyond the part-whole adaptation, there is whole evolution. To understand the creation of new species, we need to understand how the whole leaps from one state to another over time. Such a theory will, of course, not be unique to the evolution of biological species, but a more general theory of any kind of dynamics in nature. Biology and paleontology are, however, in a rather unique position to provide the data necessary to validate the ideas in such a theory.

Punctuated Equilibrium

Biologically astute readers, or those who are familiar with the debates in evolution, would have recognized by now that what I'm speaking about is similar to what is also known as Punctuated Equilibrium (PE) in evolutionary circles. PE was originally proposed by paleontologists Niles Eldredge and Stephen Jay Gould in 1972. PE proposes that most species exhibit little change for most of their geological history, remaining in stasis. Significant evolutionary changes occur rapidly and create new species not gradually but by splitting existing species into two (or more) species. The model that I will describe agrees with PE on the idea that significant evolutionary changes occur rapidly. However, my model disagrees with the idea that new species are created by splitting current species.

In the model that I will discuss, matter is semantic and its evolution is governed by the principles of idea evolution. Just as incremental additions of words to a sentence do not necessarily create a new meaningful sentence, similarly, not all random or arbitrary changes to a semantic system are meaningful. Just as splitting a sentence into two does not produce two independently meaningful sentences, similarly, breaking a semantic system does not automatically produce meaningful things. This observation about sentences refutes the conventional thesis that meanings can be created by random mutations or natural selection. In addition, when matter is treated semantically, matter is simply a representation of that meaning. The meanings or ideas themselves exist even prior to their representation or embodiment in matter. This implies that all species are eternal as ideas and possibilities. These ideas, however, appear and disappear in a cyclic fashion with the passing of time.

While PE captures some salient characteristics of the nature of evolution, it does not provide the theoretical basis for grounding these ideas into more fundamental principles in nature. For instance, PE attributes the large scale changes in the environment to weather or geographical changes, which pushes the explanation of biological evolution into the evolution of the weather or geography. This attribution is not false; however, the problem of evolution is now a problem for the geologist or the weatherman. What tools and techniques do the geologist or the weathermen have to explain the evolution of the Earth? If the weather and geology are again explained on the basis of physical theories—which at the fundamental levels are linear theories—then we would not be able to construct non-linear explanations of the evolutionary phenomena. Indeed, in current physical theories we cannot even formulate ideas about structure and function in a manner that is relevant to biology.

Therefore, while PE indeed captures some salient features about the nature of evolution, the theory is itself incompatible with fundamental physical principles embodied in modern science. PE would be a sound scientific theory only after these fundamental principles are revised to incorporate non-linearity, structure, and function. That change, I will argue, requires semantic principles.

Semantics and Biology

Semantics is, however, not just useful for explaining biological evolution. A theory of meanings is essential to talk about all everyday phenomena, including the fact that living beings have meaning processing capabilities and intentions. Following Descartes, who separated the mind from the body, science has pursued an ideology of matter separated from the mind. This in turn has made the study of meaning and intentions very hard. Materialist biologists and evolutionists in fact claim that the mind is nothing but the property of the brain which is nothing other than atoms and molecules which in turn don't have meanings. The point at which a complex aggregation of atoms and molecules becomes meanings is therefore a mystery.

To solve this mystery there is a need for natural theories in which atoms themselves are treated as symbols of meanings rather than as meaningless things. There is a need to incorporate meanings within mathematics and computing as well, to accurately describe the meaning processing capabilities in the material world, which can then be used to describe a living being's brain. The problem of meaning is not just hard, it is in fact impossible to solve in the physical view of nature, as I will detail in the next chapter. A semantic view of nature is therefore necessitated not just for explaining the evolution of species but due to even more fundamental reasons in mathematics, physics, and computing. Indeed, the laws about semantics must first arise within fundamental theories of nature before their large-scale implications can be truly understood in the everyday world. However, to formulate these laws of nature we require intuitions about semantics to be drawn from a different part of our everyday experience. Physical theories that describe nature as meaningless objects eliminate the meaning content in the symbols to arrive at descriptions in which symbols are objects. These descriptions are incomplete physically, mathematically, and computationally. Developments in biology based on these incomplete theories are also incomplete as they will not explain the nature of the mind, the sudden large-scale changes, structure, and function.

I, therefore, find it necessary to connect the problem of biology to problems in other areas of science because without this connection it is difficult to see the need for an overall shift in science. PE is a good phenomenological description of evolution, but it is grounded in current physical ideas about nature which are incapable of explaining the observed facts—especially when we widen the gamut of facts and theoretical problems that need to be addressed.

As I earlier mentioned, my previous books *Gödel's Mistake* and *Quantum Meaning* elaborate on the issue of meaning and its relation to the problems of indeterminism, incompleteness, uncertainty, and logical paradox in physics, mathematics and computing. These books discuss the application of semantics to the foundations of mathematics, computation, and atomic theory, illustrating how the problems in modern science can be solved through an explicit induction of meaning. The use of meaning in science involves a reversal of the current reductionism in which wholes are created from parts and the meaning in the whole is a byproduct of the random aggregation of parts. The reversal is that the wholes must exist in a form prior to the parts and the parts are produced from the whole by adding information. Like the ideas of 'chair leg' and 'chair seat' are produced by modifying the idea of 'chair,' similarly, the parts of the chair are produced by dividing the chair as a whole into parts.

In the physical view of nature, objects exist in a flat and open space-time. In the semantic view of nature, objects and locations in space-time are produced by adding information. The space and time in a semantic universe have to be described hierarchically, with elements of this hierarchy comprising closed boundaries. In the case of space, these boundaries will represent systems (or ecosystems in biology). In the case of time, these boundaries will represent time-cycles (or the cyclic evolution of species over geological periods). The hierarchical notions of space and time will also induct a new way of thinking about structures and functions, as these would now be defined not as a relation between parts but as a relation between part and whole (the parts being produced by dividing the whole). The next chapter discusses in greater detail the motivations for adopting these changes in physics, mathematics, and computing.

Once the problems of structure and function in biology have been connected to the problems of meaning in mathematics, physics, and computing, the solution to the biological problems would also be based on a new kind of physical theory. The form of explanation in the physical theory would also determine the type of evolutionary theory in biology. For instance, both physical and biological theories will use a hierarchical notion of space and time rather than a linear one. The hierarchical notion will change the way objects are conceived as parts of wholes rather than as independent things. This shift will in turn help us understand the unique properties in structures and functions that cannot be captured through current physical models of nature. The evolution of these structures and functions will now involve time-cycles rather than linear time.

The rationale for incorporating these changes can be seen in biology today in the problems that arise in the description of ecosystems (which include ecologies, economic systems, and organizations); however, the problems are so fundamental that theories about these phenomena cannot be developed without a profound shift in physical, mathematical, and computational theories. The second chapter therefore describes the problems of incompleteness, indeterminism, and incomputability in science and discusses their implications for biology. The third chapter discusses the solution to these problems in physics, mathematics, and computing. The fourth chapter connects that solution to an understanding of biological species, and how the species evolve under new kinds of laws. A species is now not defined as physical things but rather as meanings, because atoms and molecules are themselves viewed as symbols. The evolution of these species now follows a pattern similar to the evolution of ideas. In so far as all ideas are always theoretically possible, all species are also theoretically always possible although special conditions are needed to convert these ideas into symbols.

Accordingly, the evolution of species is not a question of how *new* species are created. It is rather a question of how ideas that are always possible become real things. An idea may exist in my mind, but it may only occasionally be verbalized into a sentence. Similarly, all species always exist as ideas, although they are only occasionally manifested. One key feature of semantics is that ideas are defined mutually or not at all. To express an idea into matter, therefore, requires the simultaneous appearance of multiple ideas that collectively reinforce and justify each other. Just as ideas exist within mutually reinforcing collections of ideas, similarly, species also exist as parts of mutually related species that form ecosystems. No individual species therefore appears independently. Rather, entire ecosystems must appear collectively and evolve collectively. This fact is consistent with observed facts about fossil records but requires a radically new theory of nature that treats matter as a symbol or representation of meaning. That theory in turn requires a revision to the current scientific thinking about matter, space, and time.

The production of new ideas is tied to the existence of minds which can create and consume meanings. While matter can encode meanings, they have to be produced and consumed by minds. This allows us to answer some profound questions about the nature of life. What do we mean by life? Is it simply the ability to self-replicate and reproduce or are meaning and intentionality essential to something being called living? These are very hard questions, especially when we try to answer them from the standpoint of current science. The understanding of these questions has further been obfuscated by a misinterpretation about the nature of matter, causality, and change based on the picture of reality painted in classical physics. For instance, if matter is meaningless then how can minds have any meaning and still be real? The problems of causality in science—when they are connected to the problem of meaning and intention—can demystify these issues. The problems will show that a world without meanings is incomplete—not just in the sense of not fulfilling the human need for meanings but in the specific sense of not predicting all empirically observable facts. These developments will also show how matter is always conserved but meanings can be created and destroyed[2]. The laws of nature are not about how matter is conserved, but how meanings are created and destroyed. Since the creation of meanings depends on minds, the evolution of meanings is the evolution of minds, and thereby that of matter.

The number of possible species is the number of possible types of minds. Unlike current biology which describes species in terms of their bodily structure, the new approach will describe species in terms of the similarities and differences between their minds. Of course, a mind can still be described behaviorally in terms of the kinds of meanings it creates and consumes. However, the 'behavior' in question is not physical tokens but rather meanings. Accordingly, the question of a species would not be based on things that we cannot observe. It would rather be based on things that we can currently observe but which we now interpret physically. When the same observations are interpreted semantically, a new way of defining species emerges from the observed patterns of behavior. This use of patterns to identify species is already prevalent in current biology although how molecules produce behavior is not well understood. A semantic approach to biology will think of a species as a mind which causes the semantic behavior to be produced. The mind cannot be observed, but its effects can be both observed and classified.

Book Summary

The questions about the stability of an ecosystem, that I earlier described, change our perspective in biology. In particular, these questions force us to construct the parts of an ecosystem from the ecosystem as a whole. In current science, complexity is built from small things to big things. In a semantic theory, complexity will be built from abstract ideas to detailed ideas. The whole will now represent abstract things and parts will be constructed by adding information to the whole. The smallest parts will be physically the most simple but semantically the most complex. Biological entities lie intermediate between the biggest and the smallest. These entities therefore have more semantic complexity than the universe as a whole (treated semantically) but less complexity relative to the atoms. Biology, in particular can also be done using everyday concepts rather than having to be constructed from the analysis of atoms, as in physics.

For instance, it would now be possible to speak about the *meaning* of words such as human being, bird, fish, and insect. These meanings would be defined through a conceptual hierarchy by which these concepts are created and the role they play in the ecosystem of living beings. They would be understood through the conceptual distinctions with other such words and through the intentional relations of give-and-take between the various species. The evolution of such species would be governed by the evolution of the ecosystems, which in the semantic view are semantically simpler. The evolution of the species cannot now be separated from the evolution of the universe; both forms of evolution will have identical theoretical models of prediction and explanation and would be connected because living beings are a subset of the entire universe.

Nature has so far been described using a bottoms-up approach where we build complex things from simple things, and the simple is defined physically. It is the goal of this book to show that nature can also be described using a top-down approach where we still construct the complex from the simple, although the simple itself is defined semantically rather than physically. The rationale and justification of this shift is, however, not limited to biology. Rather, it stems from fundamental problems in physics, mathematics, and computing. The genesis of these problems is that nature allows the potential to encode meaning in matter although science has currently described matter without meaning. When material objects are symbols, then structure and order in matter represents additional properties which cannot be reduced to the physical states of the individual objects. This is quite similar to how words in a sentence encode a more complex meaning which cannot be reduced to the shapes or sizes of the words in the sentence. To understand these properties we cannot simply add new properties of order and structure without revising our notion of the objects themselves. Rather, we must first view objects as symbols of meaning before the order and structure amongst the symbols can be seen semantically.

Symbols have contextual properties by which a symbol denotes meanings in relation to other symbols and referential properties by which a symbol refers to other symbols. To understand the contextual and referential properties we cannot just look at a single object. We must rather find new ways of describing collections such that the individuals in that collection can have contextual and referential properties. The problems of incompleteness, indeterminism, uncertainty, and logical paradoxes in science can be understood as an outcome of trying to describe symbols in terms of physical states.

When matter is described semantically, the problems of incompleteness, indeterminism, uncertainty, and logical paradoxes are resolved. This view however changes the scientific outlook about matter. Now material objects are symbols and space-time is hierarchical. This shift in thinking entails dramatic revisions to the idea of evolution itself; specifically, that evolution is not about random mutations in the parts followed by natural selection of parts. It is rather about the collective and cyclic evolution of the wholes.

A deeper appreciation of the theoretical revisions necessary in biology cannot arise without a prior understanding of the problems in computing, physics and mathematics, and it is towards an analysis of these problems that I will turn now. Unless these limitations are well-understood, their implications to biology will also remain unclear. The second chapter analyzes the problems of meaning outside biology and the third chapter will present a solution to these problems. The fourth chapter will bring the implications of the solution to these problems into the biological context. The fifth chapter will compare these shifts to the other emerging theories and viewpoints about the nature of life, and biological evolution.

2

Meta-Biological Considerations

People can foresee the future only when it coincides with their own wishes, and the most grossly obvious facts can be ignored when they are unwelcome.

—George Orwell

The Problem with Random Mutation

In the previous chapter we saw how the ideas of random mutation and natural selection cannot explain the creation of new species, even if we assume that a living ecosystem has already been formed. The problem is even more severe if random mutations and natural selection have to explain the formation of ecosystems. I also indicated a high-level view of the semantic approach that can provide a predictive theory in biology. Let us now turn towards the discussion of the problems in physics, computing, and mathematics that motivate and justify this view before we return to discussing the implications of the semantic view again for biology. I will begin with the idea of random mutations, and explore its connections to the questions of causality in nature, followed by the relation between random mutation and the problem of randomness in atomic theory.

The idea of random mutations appears to be a very straightforward hypothesis until you realize that it is not a single idea but infinitely many different ideas. As a point of contrast, it is helpful to see the difference between probability and randomness. On the face of it, both seem to imply uncertainty. However, probability involves a far greater level of certainty than randomness. Probability allows a finite[3] number of different ideas, although not arbitrary ideas. Probability fixes the relative proportions of these various ideas, and tells us which ideas are impossible. Randomness, on the other hand, allows infinitely many ideas; it does not tell us anything about the relative proportions between these ideas, and it does not indicate which ideas are impossible. For a scientific explanation, probabilities are unsatisfactory because they permit a finite number of alternatives. Randomness must be much worse because with randomness we permit every alternative and do not pick out anything.

Of course, in the case of biology, we might claim that randomness is only a euphemism for probabilities—albeit over a very large set of options. However, in this case, the problem would still reduce to why our predictions are only probabilistic and not deterministic. Is the probability an indication of missing causality? If yes, how will that missing causality be bridged to make testable predictions? A theory that uses randomness to explain biological diversity is not better than the theory that leaves the phenomena unexplained. Probabilities would be much better, but not entirely satisfactory.

I am not suggesting that mutations do not happen, but I am questioning whether these mutations are random. It may as well be that we still don't understand the causality underlying these mutations. Generally, in all fields of science, randomness is considered an undesirable feature of the theory, something that needs to be overcome by finding a better theory. Every field of science when faced with randomness seeks to find explanations for it. The belief in the order, regularity, and rationality of nature is a foundational principle of all science. To inject randomness into this ordered picture of nature seems anomalous because it essentially implies that while we can observe some facts, we cannot explain those facts. I will discuss this particular problem in evolutionary theory from various perspectives in this book. In particular, my goal is to show that the hypothesis of random chance mutation is false not because there are no mutations but because they are not random. The current hypothesis of randomness is an acknowledgement of the fact that we don't really know what is happening in nature. This is unsatisfactory as a scientific goal for any theory. A theory that explains mutations is a scientifically preferred theory than one that assumes they are random. If such an explanation requires us to discard other hypotheses in evolutionary theory then that should be easily done.

This critique of evolution is different from most other current critiques. In current critiques, there are questions whether random events persist long enough to give the mutant a significant evolutionary advantage. If the effect of a random event is short-lived and thereby reversed by the effect of a random event in the future, then random events will never accumulate to cause a significant difference to the outcome. There are also questions about the probability of random events and whether they happen frequently enough to add up to a level where they can be used to explain the enormous biological complexity and diversity. It is possible for instance that there are random mutations which cause minor changes to the genetic constitution, but not frequent enough to explain the tremendous amount of information that biological diversity requires. Then there are also doubts about whether mutations actually add information to the gene or merely destroy existing information. In the wireless transmission of digital messages, for example, there is a high likelihood that the received message is garbled and a low likelihood that the receiver hears something meaningful but totally different. Critics argue that a low likelihood of new information being added randomly implies that evolution cannot explain species.

For me, these issues about the frequency, longevity, and the effect of random chance mutations are secondary to the issue of whether the mutation itself is random. If the mutations are random, then science is incomplete because it cannot provide explanations for these facts. By calling something random, biological evolution allows a greater room for magic and miracle within science than most scientists would like to rationally accept at this time. By permitting random chance events, which cannot be predicted by any laws of nature, the 'explanation' being offered by the theory is no better than the creationist ideas that such a theory aims to debunk. I will therefore sidestep issues about the frequency, longevity, and the effect of random mutations and focus on the explanation of such mutations. If we can causally explain mutations, the issues about frequency, longevity, and effects will also be seen differently. To understand how science could construct explanations of what is currently being seen as random let's explore the ideas of randomness as they currently exist in physics, and their relation to biology.

Randomness in Physics and Chemistry

The idea of random chance mutations in biology currently has some theoretical justification in the theory of atomic phenomena that offers only statistical predictions. The idea is essentially that if atoms themselves are governed by statistical laws then by extension molecules, DNA, and the biological entity's evolution—which are all built up from atoms—must also be probabilistic. To the extent that we cannot predict the behavior of atoms, we also cannot predict the evolution of species. While this appears like a sound physical explanation of the random mutation thesis in evolutionary biology, the fact is that quantum physicists are still trying to solve the apparent conflict between the statistical nature of atomic reality and the determinism of classical physics. This is called the measurement problem in quantum theory where an uncertain world of atomic objects supposedly gives rise to a deterministic world of classical physics.

The problem is unsolved because quantum theory is fundamentally statistical and attempts to bridge this incompleteness using classical physical means have failed. Indeed, Bell's Theorem—a landmark no-go theorem in physics—proves that any attempt to make quantum theory deterministic, either by adding new variables or through new observables in matter, must lead to contradictions.

Thus, it might seem that probability is an inescapable part of reality and its use in biology as random chance events only exemplifies and extrapolates the indeterminism in physics. The only problem is that physics is itself inconsistent with one primary observed fact about the everyday world, which is that our everyday world seems to comprise relatively *stable* objects. If quantum probabilities were to be taken to apply to the everyday world, there would be no object until someone observes. Furthermore, objects would be randomly created from space-time and would dissolve back into space-time. Therefore, even if objects existed, they could not be stable. Obviously we don't see that in the everyday world. Objects seem to continue to exist even when we are not looking at them; the world does not disappear when you turn your back towards it. Nor do everyday objects randomly disappear into the vacuum, releasing energy, and neither are they randomly created from the vacuum.

Physicists take this to mean that there is some fundamental difference between atomic and macroscopic worlds. The atomic world appears and disappears randomly but the macroscopic world consists of stable objects that behave deterministically. And yet, how exactly this difference arises is not known. Quantum theory is supposed to be the theory for both macroscopic and atomic worlds but given the conflict between the probabilistic nature of quantum theory and the relatively stable nature of everyday objects, physicists selectively apply quantum principles to atoms and not to the everyday world. This problem has not been solved and, pending its solution, the idea that quantum probabilities are permanent limitations of atomic theory is false. Obviously, we know that we cannot bridge the gap between atomic theory and the macroscopic world. But it is possible that we find a new theory of atomic phenomena which is consistent with both macroscopic and atomic worlds. I will discuss such an alternative in the third chapter and even subsequently.

The randomness in current physical theories indicates that we cannot correctly predict the observations using current theories. But opinion is divided on whether this randomness is a feature of reality or a shortcoming of the theories that describe it. If randomness is indeed a feature of reality then there are so many things possible that to suppose that this randomness is somehow curtailed to produce definite life forms itself requires another explanation. If, however, randomness is not a feature of reality, but a feature of our theories, then the theoretical postulate about random chance events is incorrect and needs to be substituted by a new causal mechanism.

The idea of evolution is replaced in both these eventualities. If nature is random, then evolution cannot provide a mechanism that overcomes this randomness to produce something stable. If nature is not random, then the hypothesis of randomness is a theoretical shortcoming. Either way, the problem of randomness needs a solution in biology. However, since the problem is not unique to biology and originates in the statistical nature of atomic physics, it is likely that the solution of the problem in physics will also fix it for biology.

A further problem that non-physicists generally don't appreciate is the indeterminism in choosing a quantum wavefunction basis. What is a basis? A basis represents a specific distribution of energy within a quantum ensemble. Quantum theory formulates a Hamiltonian to compute the wavefunction, which represents the *total* energy in the ensemble. This energy can be distributed in the ensemble in many ways, each time creating a different set of objects. If you were given a certain amount of clay, you could make many different sets of objects, keeping the total amount of clay constant. Quantum theory similarly tells us that the total energy in an ensemble can be distributed in many different ways and these ways correspond to the different eigenfunction bases. Since each basis indicates a different set of particles, the idea that there is indeed an external reality that exists logically prior to observation is falsified. It is rather more correct to assume that reality is created during observation.

This reality is statistical which means that all the parts of that reality cannot be known at once. Quantum experiments distribute the same total energy into a different set of particles each time, and these particles can only be observed over time. There is hence no *a priori* fixed set of particles across different experiments performed on the same ensemble. Furthermore, even within an experiment, the particles cannot all be known simultaneously. During a specific experiment (e.g., a slit-experiment), the setup (i.e. the number of slits used in the experiment) determines the eigenfunction basis and the different possibilities in that basis are detected one after another. It is therefore possible to observe the same reality in many different ways, and the classical physical idea that observation reveals the nature of reality as it existed prior to observation is false. Furthermore, the parts of one possible description cannot all be known at once. Quantum theory seems to involve a role for the observer's choices and limitations about how much of reality can be known.

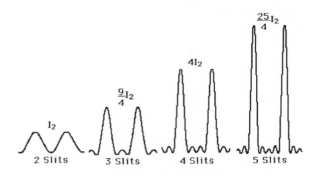

$\frac{25}{4}I_2$

$4I_2$

$\frac{9}{4}I_2$

I_2

2 Slits   3 Slits   4 Slits   5 Slits

Figure-5 A Quantum Basis Varies with Experiments

The choice of the basis and the choice of a specific alternative in that basis are different kinds of choices. The first choice implies that current theory does not *a priori* predict which events are possible. The second choice implies that the theory does not *a priori* predict the order amongst events. These two kinds of choices are incorrectly ignored when quantum theory is used to develop a theory of atoms and molecules, such as in the case of chemistry and biology, thereby a creating a picture of nature in which objects exist even when no one observes them, and they all exist simultaneously.

If chemistry is an extension of quantum physics, then there are infinitely many different sets of molecules which have the same total energy; these molecules will represent a specific *basis* in quantum theory. Furthermore, these molecules cannot exist simultaneously because each molecule has a probability of being detected at any given time. The same reasoning can also be extended to the atoms that make up the molecule. That is, the atoms are not fixed and they also don't exist simultaneously. If the atoms do not exist simultaneously, then the idea that these atoms form stable bonds with other molecules becomes problematic. For instance, in the methane molecule (CH4) the atoms of carbon and hydrogen cannot be said to exist simultaneously; each has a probability of being found at any given time, but they will always be measured one after another. If carbon and hydrogen do not exist simultaneously then the idea that they form stable bonds through interaction itself becomes suspect. Since the electrons in these atoms can only be measured one after another, the bonds must also appear and disappear one after another. This brings into question some of the fundamental assumptions in chemistry which treats molecules and bonds as physical entities.

When this problem is combined with the fact that there are many equivalent ways of distributing energy in an ensemble—which implies that even the electrons we measure are themselves dependent on the choice of eigenfunction basis—we are led to the conclusion that there isn't a fixed set of sub-atomic particles which form a set of fixed atoms which form a fixed set of molecules. Quantum theory only tells us that there is a total amount of energy which can be divided into sub-atomic particles, atoms, and molecules, and there are many ways in which this division can be performed.

Therefore the assumption in chemistry that there is indeed a single set of molecules which comprise atoms which comprise sub-atomic particles that exist simultaneously is inconsistent with quantum theory. While quantum theory says that we cannot know which molecules are real, pragmatically we suppose that there is indeed a definite set of molecules. While atomic theory says that we cannot know if all the atoms in a molecule exist simultaneously, pragmatically we suppose that they do exist simultaneously. This pragmatism is inconsistent with quantum principles but it injects into chemistry a level of classical, deterministic sanity using which chemists and biologists are able to build subsequent theories. With these two approximations, the notion of atoms and molecules as classical *things* is constructed. Now chemistry speaks about chemical reactions in a way similar to the motion of classical particles.

I am belaboring this point because the notion of atoms and molecules that form the firmament of reality—e.g., living bodies—is itself incompatible with quantum theory. This is not just a deep mystery about a conflict between atomic and macroscopic worlds. It is also a deep mystery even about the existence of chemical atoms. The pragmatism of classical physics—where we suppose that there is indeed a reality that exists independent of our observation—is applied over and over again in chemistry until we reduce indeterminism and statistics to classically certain atoms and molecules.

The pragmatism is however not necessarily false. That is, the inconsistency between quantum theory and the idea of chemical atoms and molecules does not necessarily imply that these atoms and molecules don't exist. All I am trying to say is that their existence is not explained by current quantum theory. There should be a mechanism by which a stable world of atoms and molecules emerges from the quantum world, but we still do not know what it is.

The mechanism that converts the indeterminism and statistics of the quantum world into the observed definite atoms and molecules involves choices about an eigenfunction basis and the choices of order between eigenfunctions. The mechanism by which choices create the observed definite world should also be the mechanism by which changes to this reality take place. For instance, if a molecule is a choice of energy distribution then the changes to this molecular structure is also a choice of energy redistribution in the ensemble. A chemical reaction according to quantum theory is one of the many possible ways in which energy can be redistributed, similar to how an eigenfunction basis is one of the many possible ways in which an ensemble of energy can be divided into individual particles.

This redistribution of energy is consistent with quantum theory but it represents the choice of an eigenfunction basis in the theory. If the explanation of the original set of molecules was a choice that distributed energy in a certain manner, then the chemical reaction can also be seen as a choice that redistributes the same energy in a new way. This point is important in light of claims about random chance mutations. All these mutations are chemical reactions and then can be seen as redistributing energy in some ensemble. If the creation of the original molecule involves a choice of eigenfunction basis of matter distribution, then the mutation of that molecule is also a choice about redistribution. We cannot call the redistribution random if we do not call the original molecule also as random. Both the original molecule and its mutation must be seen as redistributions of matter that preserve the total energy in the ensemble. If there is indeed a fixed set of molecules which present a definite distribution of matter, then the choice that fixed the original matter distribution must also be the mechanism to transform the distribution into a new set of molecules that we call a random mutation.

It follows that if the molecular reaction is random then the molecule is also random. We cannot suppose that there is indeed a fixed set of molecules whose reactions are random mutations. We must rather say that both the original set of molecules and their transformation are equally random. Indeed, the original and new molecules are byproducts of an eigenfunction basis selection. To the extent that this selection is a choice in current quantum theory—and hence remains unexplained—the random mutation is also unexplained. The theory that will explain the random mutation will also be the theory that explains the original choice of basis. Therefore, the idea of random mutations would not exist in a theory that explains how an ensemble of energy is distributed into particles.

The problem for biology is that it does not take the idea of randomness all the way, as quantum theory—upon which the idea of randomness in biology is based—enjoins. Rather, biology presumes a classical reality of atoms and molecules by omitting the randomness from quantum theory and then postulates a different kind of randomness that causes chemical mutations to molecules. This approach to evolution is so pervasive and fixed in biology that biologists hardly realize the conceptual mistakes being committed by their view. The problem for quantum theory is and has been the inability to show how a random reality becomes stable. When this is explained correctly, there will be no reason for supposing randomness. And until we solve this problem, the thesis of randomness is inconsistent with classical pragmatic assumptions by which randomness is reduced to the existence of atoms and molecules. Experiments clearly tell us that there are stable atoms and molecules. Therefore, the missing piece in theory is how the stability emerges. The mechanism that explains this emergence is also the mechanism that will explain what we presently call random chance mutations.

The Missing Causality in Quantum Theory

Classical physics painted a picture of causality in which forces of nature impel objects to move in space-time. These forces are byproducts of physical properties in objects, and this leads to the idea that matter moves in space-time due to its physical properties. Furthermore, all physical changes were modeled as changes in physical properties and motion of objects. This picture of causality has had an enduring impact on the rest of science, and biology is no exception. The only problem is that atomic theory changes this picture of causality in ways that most scientists outside physics do not fully appreciate. The fundamental shift in thinking created by the advent of atomic theory is that forces of nature do not cause objects to *move*. Rather, these forces put quantum objects into *stationary* states. Thus, every ensemble of objects, whose total energy is constant, comprises particles that exist in stationary states. To transform an ensemble from one stationary state to another requires a transfer of energy to or from the ensemble. However, the *mechanism* of this transfer is not explained by quantum theory. The theory only predicts that there is a certain probability of an electron jumping from one molecule to another, thereby absorbing or releasing some energy and causing a chemical reaction. But the cause of this electron jump (or that of energy transfer) is itself not explained.

Chemists hypothesize that molecules collide like billiard balls and they exchange electrons and energy in that process. But this is not what atomic theory tells us. In atomic theory, the transfer of energy and electrons has a certain *transition probability* for the quantum wavefunction to collapse into a definite state. Assume, for the moment, that the state of molecules before a chemical reaction is S1 and after the reaction is S2. Atomic theory states that for a reaction to occur there must be a combined wavefunction in which S1 and S2 (states before and after the reaction) are two different eigenfunctions. The chemical reaction takes place when the wavefunction 'collapses' from state S1 to S2. However, for the state transition (chemical reaction) to take place, the probabilities for S1 and S2 must themselves change such that S2 is now much more likely than S1 (note that before a chemical reaction S1 is more probable).

If the probability for S1 is much larger than the probability for S2, then the reaction will never occur. If the probability for S1 and S2 are comparable, then the reaction may take place but it is reversible. The only way a reaction can transpire is if S1 decreases and S2 increases. But changes in probabilities of a system require a change to the material composition and energy of a system, and to effect this change, a second system is required to exchange matter and energy from the first system. However, the second system is also governed by quantum theory and has the same predicament—i.e. the second system can only change its probabilities if some other system exchanges matter or energy with it. The changes to the second system therefore require another third system and so on.µ

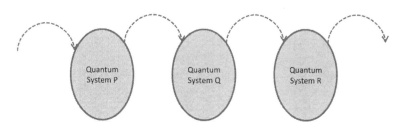

Figure-6 The Problem of Quantum Causality

Quantum theory creates a problem of causality that scientists outside physics do not appreciate very well. The problem is that every system is in a stationary state if its energy and matter are constant. This system will not automatically emit or absorb energy and matter unless impelled to do so by another system. However, if every system is in a stationary state then no changes can occur.

Chemical reactions involve a variation of the famous Zeno's Paradox in which before an object can go from state A to B, it must go half-way to that state, and before it can go half-way, it must go one-fourth of the way, ad infinitum. Classical physics solved this problem by asserting that an object in motion has a momentum due to which it moves at a constant speed regardless of its position. This assumption is false in atomic theory because when an object has some fixed momentum, it is in a stationary state and not in a state of motion. If objects are in a stationary state, then to change that state something must change its own state prior. But for that cause to change state there must be another cause which changes its state prior to that, ad infinitum. Greek philosophers were acutely aware of this problem and Aristotle suggested that for change to be initiated there must be an Unmoved Mover which causes other things to move but is itself unmoved. The problem of original cause remained unsolved for nearly two millennia until Newton shifted the focus from seeking the original mover to studying *changes* to motion.

The essential import of Newton's idea is that we cannot answer questions about the original mover. Rather, since we observe the universe already in motion, we can take motion for granted and then study the changes to the states of motion. Newton's first law of motion therefore states that a body continues in motion unless disturbed by a force. The second law computes the changes to motion when a force acts. To create this model of change, Newton had to make a crucial assumption—namely that an object in motion has a momentum and an object that has a momentum must be in motion. Quantum theory invalidates this crucial assumption because now a particle has a momentum but it is still in a fixed position state.

In Newton's physics, objects have to be set into motion one time, after which they would always be in motion. In quantum theory, every state transition requires an Unmoved Mover because every state—if the energy is constant—is a stationary state. Chemical reactions are quantum state transitions and they can occur only if the probability for the chemical products exceeds the probability of the chemical reactants. This change in probabilities itself requires a state transition, which in turn requires another transition, etc.

Quantum theory does not explain change in the way that classical physics did. Unlike classical physics which depicted states of motion, quantum theory depicts stationary states. The state changes if there is a discrete energy transfer, but the cause of that transfer is outside the theory. The problem of missing causality in physics is seldom understood by chemists and biologists who take the idea of motion and change for granted. They assume that objects are already moving; that they collide, which results in chemical reactions, which is the root mechanism for biological evolution and the emergence of living species. They also assume that physics explains how material objects move in space-time which is not entirely true.

Again, I am not insisting that there are no chemical reactions, but only pointing out that current quantum *theory* does not explain how these reactions take place. Quantum theory only states that an electron will jump from one state to another when the energy (or force field) in the ensemble is changing. But for that energy to change, another ensemble must emit or absorb energy, and since that ensemble is also in a stationary state unless its own field is changing, there must be yet another system that emits or absorbs energy to cause a change in its field, and so on, ad infinitum. If a single transition cannot be causally explained in current physics, then the succession of such changes is also unexplained. Biological functions are examples of sequences of hundreds if not thousands of chemical reactions, none of which can be explained in physics.

Note that this problem is not how the chemicals were assembled together to form certain structures through evolution. The problem is rather that *given* those structures were already formed there are still explanatory gaps in the causal explanation of how the biological machinery actually works within living bodies. Obviously, this makes the question of how the molecules were assembled harder. Biology assumes that chemical reactions are causally explained and then extends this to random reactions followed by environmental selection. If, however, a chemical reaction itself is not causally explained then its extensions into mutation and selection to form a theory of evolution of biological function is also incomplete.

If, however, there were a theory that explained how a chemical reaction takes place, in a way that did not involve missing pieces of causality, then such a theory would not require the notion of random chance mutations. If a chemical reaction has a causal explanation then it cannot be random. Even the idea that a DNA is mutated randomly because of radiation (such as due to sunlight) itself requires an explanation of chemical reactions in the Sun that emit the energy. But how chemical (or nuclear) reactions are taking place in the Sun is itself a problem of causality unexplained by present quantum theory. Furthermore, even if we assume that the DNA is mutated by the radiation, the DNA still has to be transcribed in the cell to produce proteins, and this transcription also requires chemical reactions. Quantum theory only explains that given a certain amount of energy, the energy can be distributed into particles in infinite possible ways. It does not explain how the energy goes from one system to another. Therefore, if the universe was already formed into molecules, how these molecules react cannot be explained.

The problem in physics today is that it is a theory of matter but not a causally complete theory of change. Of course, the problem of causality is not new, because it existed even in Greek times in the need for an Unmoved Mover. Classical physics shifted the Unmoved Mover to a one-time action of something that sets the universe in motion, after which the universe remains in perpetual motion. This idea of motion is now known to be false. It has been replaced with the idea that force and energy put matter into stationary states, not into states of motion. To cause motion, the energy must change. And this change is not about an Unmoved Mover at the beginning of the universe. It is rather about every state change transition in matter.

The quantum problem of causality is consistent with our everyday notions of causality but inconsistent with classical determinism. For instance, the everyday notion of causality tells us that *if* we consume an analgesic we would be relieved of pain. The existence of the analgesic does not grant pain relief because there is still a choice of consuming the analgesic. Classical determinists suppose that this gap in causality which appears as choice is an illusion because the choice is in turn a byproduct of some chemical reaction in the brain which causes the hand to consume the analgesic resulting in further reactions that relieve the pain. In classical determinism, therefore, there is actually no choice, because there are always some moving objects which cause other objects to move deterministically. Quantum theory changes this view of causality. In the quantum theory, each system is in a stationary state, not in a state of motion. The hand thus cannot pick up the analgesic unless there is a change in the force field in the brain or the hand that causes that movement.

To change any system, some other system must emit energy and this emission cannot be causally predicted. Each system has the potentiality to emit energy but when and why that energy is emitted cannot be predicted. The inability to predict the energy emission implies that the causal mechanism underlying it is unknown. And without such a mechanism, we can only suppose that the emission in one system is caused by an emission in another, ad infinitum.

The problem of causality in quantum theory tells us that we do not yet have a complete theory of natural causation, not just at the level of atoms and molecules but also at the level of macroscopic objects. The deterministic notions of cause as force only put a system into a stationary state; they don't cause the system to change or evolve. The evolution of a system is one of the several possible states, and the cause must choose from amongst the states. Without this choice, the next state of a system is underdetermined. The choices are therefore not random as they are constrained by what is possible. But the possibilities themselves do not determine the outcome as they have to be converted to actuality and that conversion necessitates a choice. This conversion therefore requires an extra causal element missing in current science. John von Neumann postulated that this choice could be the effect of the mind although the interaction between mind and matter has never been explained. Furthermore, the idea that conscious choices can provide the missing causality in science is flawed because even in a living body many processes—such as blood circulation, immunity, cell regeneration, digestion, and breathing—happen without conscious intervention although the conscious states do influence their functioning.

The complete causal theory of nature requires a type of causality that lies between current theories of matter and consciousness. It cannot be material objects because the problem of causality cannot be solved by adding new types of variables or observables to quantum theory. It cannot be consciousness because many things in matter evolve without a conscious intervention. There is a need for a cause that selects from possibilities in matter. This cause can be provided by the idea of meanings, as another type of matter different from material objects. While meanings can be embodied in matter and can influence material objects, meanings are logically prior to these objects and cannot therefore be reduced to matter. Meanings lie intermediate between matter and consciousness. The conversion of meanings into symbols of meanings is the missing causal element in science and I will later discuss how a theory that encodes meanings in space-time can solve the problem of missing causality.

In this theory, the choices represent information. The same total energy can encode various types of information; while the energy is conserved, information can be created and destroyed. The laws of energy are incomplete because the same total energy underdetermines the information encoded by that energy. Information however completely determines the energy distribution. Therefore, if the laws of nature were formulated as laws of the evolution of information, then the uncertainty in the next state would be overcome. What appears as choice and probability in current quantum theory—and therefore requires an infinite cascade of preceding physical actions—would be determined by laws of information. While the missing causality cannot be bridged in the physical view of nature, it can be resolved in an informational theory of nature.

Thermodynamic Uncertainty and Reduction

In classical physics, an object has physical properties which create forces of nature which in turn govern their motion. For instance, in Newton's physics, matter possesses the property of mass which creates a gravitational force which causes objects to move in space-time. Newton's law of gravitation specifies the amount of acceleration an object will undergo if the masses of all the objects involved are known. If this is a true picture of reality, then every object must move according to deterministic laws of nature, and the universe as a whole must be governed by these deterministic laws. After all, what is the universe if not a collection of the independent parts?

The idea that the universe is nothing but a collection of independent parts had tremendous successes in explaining the motion of celestial and terrestrial objects, as long as the total number of objects involved in the causal interaction was very small. This idea, however, ran into difficulties when physics tried to describe phenomena concerning the thermal behavior of gases, which are comprised of a very large number of molecules. It was noticed, for example, that the total amount of heat that can be *extracted* from a collection of particles is not equal to the total amount of heat that was *added* to it. Newton's physics held the conservation of energy as a basic law of nature and, therefore, the total amount of energy had to be the same whether or not we could extract that energy from a system. This was further confirmed by the fact that although we are not able to extract the energy from the system, the system still has a higher temperature (even though we cannot extract the energy for work, we can still extract it in relation to a measuring instrument—thermometer). Therefore, it was natural to suppose that the system has energy but that we cannot extract it. One way to theoretically explain this problem was to suggest that the reason we cannot extract the energy from a system has something to do with how that energy is *distributed* inside that system. For instance, if all particles in the system are aligned and move in a single direction then the total amount of extractable energy equals the total amount of energy. However, if these particles are randomly moving in different directions, then much of that energy is trapped in the interaction between the particles and it cannot be extracted from the system.

We could now call a system that has all its particles moving in one direction as a highly 'ordered' system as compared to other 'disordered' systems where particles move in random directions. Note that in the description of individual particles there is no notion about order or disorder. All particles are essentially ordered because they behave according to classical deterministic laws of motion. But when we collect such particles, we are no longer looking at each particle individually. We are rather looking at all of them collectively and the notion of order and disorder emerges through a relative comparison between the states of motion of the various particles. The order or disorder is now not a property of the individual particles, but becomes a property of the collection as a whole. The idea that a whole is nothing but a collection of particles is now false because the notion of order or disorder cannot be associated with the individual particles, and can only be tied to the collection.

While this reasoning is physically intuitive, it suffers from a mathematical problem—the laws of classical mechanics are *reversible* in time. So, if we begin with an ordered system, and find that over time it becomes disordered, there is nothing in the natural laws that prevents that system from becoming ordered in the future. If, therefore, the amount of energy extractable from the system depends on the total order or disorder in the system, then the extractable energy could change in time as the relative amount of order increases or decreases. This idea runs counter to observations.

It is observed that if an isolated system has some energy, the amount of extractable energy in that system never increases. If the extractable energy depended on the order in the system, and the order improves with time, the extractable energy must increase as well. This is obviously not what is observationally confirmed. Furthermore, it is practically impossible to compute the real particle states of motion inside a large collection of particles because the total number of equations that need to be simultaneously computed is far beyond the capabilities of any known computer. Therefore, the laws of physics cannot predict whether all the particles inside the system are perfectly aligned and moving in a single direction or randomly moving in different directions. While the idea of order and disorder can be intuitively stated, it is practically useless.

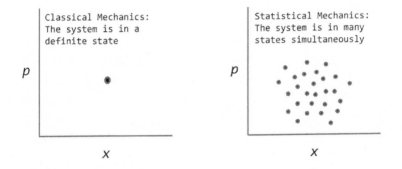

Figure-7 Classical and Statistical Descriptions

A new approach was now adopted in which the order or disorder was defined relative to the total number of *possible* states of the system as a whole. This idea is intuitively known to us as the possibility of one of the six faces of a dice turning up on a dice roll. Generally speaking, there are only six possible outcomes of a dice roll (unless the dice stands on one of the edges, which is highly improbable). The total amount of uncertainty in a dice equals the total number of faces that could turn up on a dice roll; the uncertainty equals the total number of *possible* states of the dice. Similarly, if the particles in a system *could* be in some state then these possible states would represent a measure of the total uncertainty in the system state. The collection of particles was now described in terms of its *possible* states rather than its actual states. This description is based on the simplifying assumption that if the number of possible states in a system is very large, and it is practically impossible to predict which amongst those states is real, then we might assume that the system is in any of these states, with a finite probability. This theory came to be known as classical *statistical* mechanics and it was successful in explaining the irreversible phenomena of heat.

Now the statistical treatment of particles leads to a serious problem about whether we treat the statistics epistemically or ontologically. Should we say that nature is in fact in a definite state but we don't *know* which state it is in? Or should we say that nature *is* in fact in many different states simultaneously? If indeed the system is in a single state and statistics only pertains to our ignorance of that state then the predictions of the statistical description must be false. If, however, the statistical description produces correct predictions, then the total number of possible states has a *real* role in nature and the successes of the theory would compel us to suppose that the system is indeed in an uncertain state. Classically speaking, we would have liked to believe that the system is in a single state although we might not know what that state is. However, given the predictive correctness of the statistical approach, we are compelled to suppose that the system is in fact in a classically uncertain state.

How can a classical system be in an uncertain state? Classical physics does not permit uncertain states. Classically, the system is in a definite state, although it may be hard to predict it. Statistically, our inability to predict the state has empirical consequences. Are we to suppose that the system somehow knows that we cannot predict its state and therefore it behaves as if it is in an uncertain state? Or should we suppose that the system is indeed in an uncertain state, although classically we think that all systems must be in a fixed state? Some remarkable things happen when we shift our focus from individual objects to collections of such objects. The remarkable fact about a collection here is that individually all objects are in a definite state, but collectively they are in an uncertain state.

How can a system have individual particles in definite states while the total system is in an uncertain state? This is an unsolved problem in classical physics. The problem is well-known as the irreversibility of thermal phenomena where systems appear to go from order to disorder and classically speaking we cannot even define disorder at the individual particle level. The natural conclusion from this problem is that the theory that describes ensembles or collections must be different from the theory that describes individual particles. The world that appears classically deterministic becomes statistical for collections of particles. This transition from determinism to statistics represents a gap in explanation and requires a new type of causal theory, even at the level of macroscopic objects.

The main reason why I discuss statistical mechanics is to show that classical physics is causally incomplete even with regard to macroscopic objects, when these objects represent ensembles of particles. Classical physics is said to fail in regard to the extremely-large and the super-small. This is by no means false. But what is often ignored in that picture is that classical physics also fails for thermal phenomena which are neither extremely-large nor super-small. The root cause of this failure is still unknown. But what we can say is that even if science had not been revolutionized by later physical theories like quantum mechanics and general relativity, the classical physical picture of reality would still be incomplete.

Biological systems are neither extremely-large nor super-small. They are already known to be in an uncertain state due to quantum theory. However, even if we assume that the quantum uncertainty is somehow overcome to create classical systems at some level of organizational complexity, then the classical ensembles are uncertain, too! In fact, this uncertainty has empirical consequences (non-linearity) that cannot be predicted by a classical particle theory. We now have a problem at hand: if the quantum system becomes classical then it is classically uncertain; otherwise it is quantum mechanically uncertain. Each classically uncertain state represents a distribution of matter inside the ensemble and while classically we would like to suppose that we don't know the distribution, the predictive correctness of the statistical approach suggests that matter is in fact not distributed. The only correct way to understand statistical predictive correctness is to realize that we can speak about the total amount of energy in the ensemble but we can't speak about the particles that were classically supposed to make up the total energy.

The inability to predict an actual distribution of energy in the ensemble is a problem of computing and measuring the actual state. But the fact that the statistical approach predictions match observations implies that we don't have a problem of computing or measuring the state; rather, we have the even more profound problem that the ensemble isn't in any definite state. Each such definite state would represent a specific matter distribution and we cannot suppose ignorance about a particular distribution. Rather, we must suppose that matter is not distributed, or that there is energy in the system but it is not individuated into a specific set of particles.

This is a far more profound consequence of statistical mechanics than classical physics generally acknowledges. We generally suppose that there is indeed a set of classical particles—molecules of gas or liquid—whose state is uncertain to *us*. The more accurate interpretation of statistical mechanics is that there isn't a set of molecules into which the total energy is divided, unless a measurement is performed. When we perform a measurement, the act of measuring itself requires the addition of information that in turn creates particles. The fact that we see particles upon measurement—and which in turn implies a definite state—is not how reality exists prior to measurement. Rather, measurement creates a classically definite state, but the state is otherwise classically uncertain.

The problem of knowing the exact matter distribution in statistical mechanics is conceptually similar to the choice of an eigenfunction basis in quantum theory. Both problems lead us to the same conclusion, namely that the material particles are not real until a measurement is performed. The type of measurement rather provides the necessary information to 'create' a set of particles. This creation is called an eigenfunction choice followed by a 'collapse' in quantum theory and it would be called a particle in classical physics. There are many ways in which to choose a basis or create particles, and the observed result is an outcome of an experimental choice.

This fact has remarkable consequences to the idea of random mutations because the notion of a mutation assumes the existence of molecules as *a priori* real objects, which are just being modified by a mutation. Statistical mechanics tells us that the system has only energy, but no particles in a definite state. Particles are only outcomes of our measurement on the energy, and they are produced as a consequence of the information added by the measurement.

In effect, the ensemble is like a bottle of ink and the measurement spreads that ink into a specific kind of text. Until the ink has been spread into text, the ink is in an uncertain state of being many different texts.

However, to assume that the ink is simultaneously the combination of all the texts is mistaken. It is more appropriate to suppose that there is ink although no text. The idea that ink is in the superposed state of all the texts is something we postulate to lend credence to the classical physical idea about *a priori* real particles. In reality, we have to speak about ink without speaking about texts. That in turn implies that the conversion of ink into text requires the addition of information without which the ink is in an uncertain state. If we assume classically definite particles, then the uncertainty is insurmountable. That would in turn mean that we could not observe any particles in nature because they are simultaneously everywhere. But if we assume that energy is real and particles are created from that energy by adding information, then the state is uncertain only until the point we observe the world. When we observe, we see the particles, and the state is indeed definite. When we don't observe, the state is uncertain not because we cannot predict it but because there is in fact no particle to be known.

This viewpoint equates the statistical uncertainty to missing information. A classical statistical ensemble is not just epistemically uncertain but also ontologically uncertain. Our observations provide the information that fixes both epistemic and ontological uncertainty. That in turn implies that information must be real and objective. With the information the system is classically definite. Without that information the system is classically statistical. Since definite and statistical systems produce different predictions (reversible and irreversible phenomena) information has a real role in nature. The role is simply that energy can be distributed into material particles in many ways, which cannot be predicted upfront. The exact distribution of energy into particles represents a system uncertainty, and that uncertainty leads to non-linear behavior in thermodynamics. The extent of uncertainty can also be equated with the amount of information that can be encoded into the thermodynamic system.

Non-linearity in classical physics entails the inability to separate the system into individual parts. The cause A depends on the effect B, which in turn depends on A, creating a circular causality. These causal loops in material systems imply that we cannot distinguish between the parts. Only linear systems are totally distinguishable; non-linear systems are not distinguishable. Classical physics is linear and distinguishable but statistical mechanics is not, which creates a difficulty in understanding macroscopic systems at the level of classical physics. The difficulty arises because we would like to think of the world as individual particles governed by linear theories, but experiment shows that the system is non-linear. The only possible resolution to this problem is to discard the idea that the system is *a priori* divided into parts which are somehow deeply entangled due to non-linearity. The resolution would entail that there is a system although it is not built up of parts; that there is energy but that energy is not distributed into individual particles.

A mutation in such a system is a redistribution of energy into parts. In fact, every change to such a system is a new way of distributing energy. This redistribution creates a classical system in which particles have a fixed state. The transition from energy to particles therefore entails a gap in information in which the undifferentiated becomes differentiated. A differentiated system is always linear, and an undifferentiated system is non-linear. The non-linearity in statistical mechanics is currently explained by injecting a non-physical principle of statistics into a classically definite theory, creating paradoxes of interpreting and understanding classical systems. A better approach requires discarding the classical idea of real particles and replacing it with the idea of information that creates particles.

In the previous chapter I described how a non-linear system must oscillate violently (if the causal loop represents positive feedback) or remain in an attractor (if the causal loop represents negative feedback). Effectively, a negative feedback loop dampens individual changes in the parts that occur without a concerted change in the whole. Such a system is linear at the level of the whole *and* at the level of the parts. With the understanding from statistical mechanics, we can conclude that a chaotic system is non-linear not because of feedback loops but because the system is not yet differentiated. A fully differentiated system is always linear and non-linearity is an outcome not of feedback loops but of lack of differentiation. In effect, when the particles are defined, they are not defined independently; they are always defined in relation to the collection.

This approach can help us demystify the problematic relation between statistical and quantum theories. Quantum theory is linear[4] and statistical mechanics is non-linear. Both theories are about the behavior of an ensemble of particles—i.e. macroscopic systems. Why is it that the two theories describe the same ensemble in two different ways, when each theory is empirically confirmed separately? The mystery further deepens when we realize that statistical mechanics is in fact the correct description of thermodynamic ensembles, and quantum theory is not (due to being linear). Quantum theory would not describe the non-linear behavior of ensembles. Indeed, this mystery represents an unsolved problem in physics where a fundamental theory of nature (quantum theory) is linear although the ensembles of atomic particles behave non-linearly.

The conflicting descriptions of quantum and statistical theories can be reconciled in a new theory based on the idea of information rather than *a priori* real particles. The reconciliation would be that if a system has maximal information then it is linear; otherwise it is non-linear. Quantum ensembles will now represent the state of a system which is maximally differentiated, since the theory describes subatomic particles. Statistical ensembles will represent the state of the same system when it is less than fully differentiated. Information cannot be added into a quantum ensemble, and that is why it is linear. Information can be added into a statistical ensemble and that is why it is non-linear. Non-linearity therefore does not represent some mysterious property of an ensemble of *a priori* real particles. It can rather be the outcome of the fact that the system can hold more information and can therefore be further divided. Quantum theory represents the limits of dividing an ensemble into parts by defining the smallest particles that can be created in nature.

The extent to which the energy in the ensemble is differentiated and the manner in which it is differentiated represents a choice in the current theory and it will represent information in the new theory. In the case of biology, this would mean that the non-linear behavior of biological thermodynamic systems is an outcome of the fact that these systems are in fact *not* built up of *a priori* real atoms. Biological beings are in fact real at less than maximal information. That is, it is not necessary to speak of a biological system only at the level of sub-atomic particles. In fact, if a biological system were reducible to the sub-atomic particles then it will not exhibit non-linear thermodynamic properties. Rather, we can speak about biological systems in terms of cells, organs, and functions, which are descriptions of the system in states of less than maximal information. The reductionism of current biology would now contradict the thermodynamic irreversibility of biological systems. The fact that biological systems are born, grow, become old, and eventually die would imply that reductionism is false. Reductionism would rather imply that the system is reversible and a biological being could change in reverse—i.e. from death, to old age, to youth, and then to birth.

The reversibility of fundamental physical theories and the observed irreversibility of macroscopic phenomena (biological beings included) represents the essential paradox of reductionism. If reductionism is true, then life is reversible. If life is irreversible then reductionism is false. Problems in statistical mechanics illustrate the connection between reduction and reversibility (or linearity); they demonstrate that the only correct way to think of macroscopic non-linearity is to discard the idea of physical reductionism and embrace physical divisibility in nature. That is, nature is not *a priori* divided although it can be divided up to a natural limit. If nature is maximally divided, then systems are linear. Otherwise, they are non-linear. In both cases, the whole is more real than the parts because the parts are constructed by dividing the whole through information.

A living being, now, does not begin in sub-atomic particles. Rather, a living being appears as a whole before it is divided into parts through increasing addition of information that details and elaborates the whole. Like an essay is developed from a précis, a paper is developed from an abstract, or a book is developed from a table of contents, the material body of a living being is developed from an abstract and concise representation of the idea of that being, by adding ever more structure, detail, elaboration, or information to the précis, abstract, or table of contents. The précis is also material, in as much as the essay is material. However, the précis is more concise relative to the essay. The total amount of meaning in the précis is semantically identical to the meaning in the essay, although the essay has a greater amount of physical complexity. As the précis becomes shorter, it becomes more non-linear: we can either understand it as a whole or not at all. As the essay becomes longer, it becomes more linear: different parts of the essay can be understood independently of the other parts. A system becomes completely linear when no further information can be added into the system.

The problems of non-linearity, irreversibility, and statistics in particle collections indicate a view contrary to reductionism in biology. The whole does not develop from the parts. Rather, the contingent develops from the abstract. This shift not only requires us to move away from reductionism but also away from physicalism. To understand how the contingent develops from the abstract, we have to revise the understanding of matter from things to meanings.

Choices in General Relativity

General Relativity was created by unifying two classically deterministic theories—Newton's theory of gravitation and Maxwell's theory of electromagnetism. So, it might seem counterintuitive to claim that the theory involves indeterminism. The element of surprise in this case is also higher because the indeterminism in relativity is not as well-understood as in quantum and statistical mechanics.

The indeterminism in relativity is not conceptually different from that in statistical mechanics and quantum theory. Relativity is a deterministic theory when the individual particles and their states have been fixed. By fixing the individual particles and their states, the total energy of the universe is also fixed. When Newton described gravitation, he assumed the existence of individual particles, their masses, and their initial states of position and momentum. However, when Einstein created the general theory of relativity, he began with the idea that there are space-time events and energy needs to be distributed over these events. In classical physics, we begin by assuming particles and their energy. In quantum theory we assume total energy which is used to compute particles. In relativity we assume energy and *events*. The indeterminism in relativity is based on different assumptions, but it is conceptually similar to the problems of matter distribution in statistical and quantum theories. Relativity assumes events (which quantum theory does not), but does not assume real particles as classical physics does. The indeterminism in relativity is that there can be different material particles, although the total energy and the events are the same. The determinism of classical physics is missing in relativity, although the theory is more deterministic than quantum theory because it assumes the real existence of space-time events in the universe.

The goal of Newton's gravitational theory was to describe the motion of objects, not the origin of these objects. However, Einstein showed through general relativity that matter is produced from space-time curvature. The primary reality in relativity is therefore space-time. But how do we define space-time? To create objects from space-time curvature, space-time had to be defined as events. For a given total amount of energy, there can be many space-time structures and events. In classical physics, this was known as the inability to determine the initial and boundary conditions of a system. This indeterminism exists in general relativity as well. They difference, however, is that Newton's mechanics did not aim to find the origin of motion, but only changes to motion[5]. This isn't the case with relativity and cosmologists have used the theory to determine the origin of the universe. If the same energy permits infinitely many different universes, then the theory is also incomplete.

However, even if we fix the initial states of material objects, there are still infinitely many possible redistributions of matter, which can be effected by the splitting and coalescing of particles, because such changes still preserve the total energy and momentum of the system and are therefore allowed by any mechanical theory. This problem can be solved by assuming that the total number of particles (and their properties) is fixed. In other words, particles do not coalesce and split; the particle identities remain immutable. This postulate is unphysical because physics does not deal in particle identities, and particles are not supposed to be conserved. However, without it, relativity (like classical physics) is indeterministic.

Nevertheless, even if we grant that the total number of particles is conserved, there is still a problem of *exchange symmetry* between these particles. To illustrate this problem, let's suppose that there are two trajectories with identical masses—e.g., 65 kg. These masses are in different locations and they are materially different particles. However, we can think of matter redistributions that swap these objects.

The swap will not cause any detectable change from the standpoint of gravity because the swapped masses are identical. Such a swap, however, can occur during object collisions where some of the mass of the particles is exchanged. How much of the mass is exchanged during the collision makes the theory indeterministic, because this exchange cannot be predicted, although this swap will be totally undetectable. To detect this swap we would have to assume that the swapping objects are also observers.

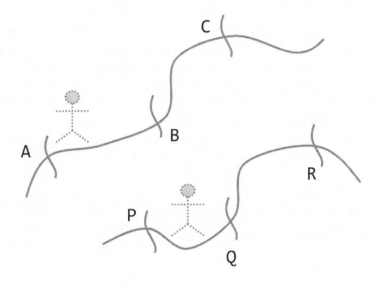

Figure-8 Matter Swapping Creates Different Experiences

If the swapping objects were different *observers*, then their swap will preserve all invariants but it will change experiences. If the first object passes through events A, B, and C while the second object passes through events P, Q, and R, by swapping the objects, the first object will go through events P, Q, and R while the second object will go through the events A, B, and C. From the standpoint of general relativity, nothing has changed. The total mass, space-time metric, and point coincidences are all the same. But, from the standpoint of the observers, the events they experience have changed.

Current relativity looks at the space-time event manifold from the 'outside,' because when identical masses are swapped, there is no observable difference if the observer is outside the universe. But if the objects in the universe are themselves observers then swapping matter in between events would be consistent with relativity but indeteterministic observationally. Through successive postulates of initial condition and fixed number of particles, we can fix the events but not the *identities* of objects passing through those events. If the observer is outside the universe, then the events equal the experience, regardless of which particle passes through the events. But if the observer is in the universe as one of the particles, then the events are different from the view of those observers. Now the fact that each event is also an experience and matter can be swapped between these events itself represents a change in experience without a change in the events themselves. Matter swapping transforms now represent choices of events that an observer passes through. Relativity would now imply that observers cannot change the events in the universe although they can change their experiences.

General relativity is *what* deterministic but *who* indeterministic. That is, it predicts what will happen in the universe but not who will experience it. The theory permits many possible matter distributions subject to the total energy, matter, particles, and events remaining conserved, because the theory is symmetric in the swapping of particles with identical properties—two objects with the same mass can be swapped without any observable difference from the standpoint of relativity. The swap will, however, create a difference to the swapped objects, if these objects were observers.

To solve this indeterminism there is a need for theories that can speak about *object identities*. Both classical physics and general relativity are matter preserving but not identity preserving. When two objects with the same mass collide, and the recoiling objects after the collision have identical masses, there is no way to know if the masses during the collision were exchanged. The problem of general relativity is that it does not have a good definition of object *identity*. We can only measure physical properties and physics is defined around physical properties, not identities. The identities in relativity are trajectories, and not objects. Classically, we assume that an object is a trajectory, but this is not necessary. The trajectory is actually a path that can be traversed by many different objects. It is more appropriate to think of trajectories as 'roles' that can be occupied by many 'actors' rather than the actions of a fixed actor. This gives an intuitive definition of matter redistribution: the redistribution is the change in which a different actor performs the same role. To perform the role, the actor must 'fit' into the role requirements. However, since nature permits multiple matter redistributions, it permits the possibility that many actors can fit the same role.

The determinism of trajectories and events (assuming that object identities are fixed) amounts to the idea that the roles and the events are both fixed. This is similar to a drama in which the script—which defines the roles and events enacted on the stage—are already predetermined by the script. The indeterminism of relativity is, however, the fact that the same role and the same events can be enacted by different actors or participants. This scenario indicates the existence of a new kind of indeterminism in which events are fixed but the participants are not. If observers X and Y swap events M and N (without changing the total mass and energy), then there is no empirical difference from a third-person perspective. But from a first-person perspective, there is a big difference.

Note how Newton's mechanics—which deals in the dynamics of individual objects—is completely deterministic because both the particle and the trajectory are fixed. Relativity becomes indeterministic because the theory deals in particle collections. There are many equivalent ways to distribute matter and energy over the same trajectories and events, which cannot be fixed by the theory. This indeterminism is not necessarily an effect observable only at galactic scales. It is also an effect that could be observed at the level of everyday macroscopic objects as well as atomic objects. Correspondingly, the relativistic indeterminism has implications for everyday and atomic objects as well, even though the effects of gravity only make an observable difference at very large space-time distances.

Matter redistributions of the above kind reinstate into a classically deterministic theory the importance of a first-person viewpoint because the third-person view is predictively incomplete.

The biological counterpart of this problem is that there many different objects which can perform the same functions. For instance, the same molecule of hemoglobin can carry oxygen to the hand or to the leg. How do we determine which molecule of hemoglobin goes to the hand instead of the leg? The same neurotransmitter can carry the signal of pain from the hand or from the leg. How does the brain know that the neurotransmitter is arriving from the hand or the leg? The transmission of oxygen or neural signal represents a trajectory. This trajectory is occupied by a specific object. Classical physics is indeterministic about which object occupies which trajectory. Classical physics only fixes the *type* of the object—defined by its physical properties—but not its specific *instance.*

For example, if you buy a paperback version of the *The Origin of Species* by Charles Darwin from some online retailer, the retailer is only obliged to give you a copy of the book, but you cannot determine which specific instance of that book you will get. There is potentially room for allowing the fact that the retailer could have shipped you a different copy of the same book. All copies have the same information (they are conceptually the same book) although they are physically different instances of the book. To ship one copy versus another, the retailer makes a decision or choice. This choice is often based on an arbitration that is beyond the buyer's control.

The indeterminism of relativity is similar: the theory indicates which type of object will be on which trajectory, but not which instance of that type. Of course, to understand this indeterminism, we need to inject the distinction between types and instances in science. Philosophically, this distinction exists in physics as the difference between a particle and its properties; the properties represent the *type* of the particle (e.g., its mass) while the particle itself denotes the *identity*. The problem, however, is that classical physics does not invoke a role for particles aside from their properties. Thus, it doesn't matter which particle is on which trajectory. It only matters that the properties of the particle are consistent with the form of the trajectory. The idea of a particle is therefore completely superfluous to the predictive and causal framework in physics. We only postulate particles because it helps us give a realistic interpretation to science: we suppose that there is indeed some object on which the properties 'hang.' This assumption is necessary when particles have multiple properties—e.g., mass and charge—because when a particle has many properties then we need something beyond these properties to tie these to the same object instance.

Indeed, when a particle has more than one property, it is possible to construct matter redistributions that clearly demonstrate the need for particle identities. For instance, if particles have both mass and charge, then it is possible to have two distributions in which the same trajectory is followed by two different particles—the first with mass X and charge Y, and the second with mass P and charge Q.

The problem of indeterminism in classical theories raises a basic question: Should we describe nature as objects (reality) or events (phenomena)? Events are empirical and constitute what makes a theory true. Objects are what we presume explains these observations. The indeterminism of classical theories establishes the following well-known philosophical fact in the context of science: facts underdetermine objects or phenomena underdetermine reality. The difference between various realities cannot be determined at the level of facts, if we try to derive the realities from the facts. The only way to know the reality would be to experience it from a first-person perspective rather than a third-person perspective. The third-person perspective only tells us what *that* is, but not what am *I*. The indeterminism indicates that the third-person perspective will always underspecify the reality and that reality can only be known in a first-person perspective. The specific identity of an object cannot be known from a third-person perspective, but it must be known from a first-person viewpoint. The choice of which amongst several possible trajectories an object must take cannot be predicted by phenomenal theories; it must be individually made.

Evolutionists have always held that there is no room for intention and choice in nature and nature is, in a sense, driven by blind chance, not an intention or a goal. That a living being mutates randomly and it is selected by the environment. And that nature is causally complete without a conscious intervention; rather, conscious interventions are themselves byproducts of material combinations. The problems of indeterminism in classical physics—which happens to be the most deterministic theory in current science—show that nature is causally incomplete and that the theories permit many possible alternatives which cannot be selected by observation. The only way these alternatives could be distinguished is if there is a sense of *identity* or individuality underlying each physical property, and this identity can choose between the alternatives.

The Matter Distribution Problem

So far in this chapter we have looked at three kinds of indeterminism problems in scientific theories. Each of these problems arises when we attempt to describe ensembles or collections of objects rather than individual objects. Even classically deterministic theories become indeterministic when dealing with object collections. Quantum theory is indeterministic and its problems with probability are legend; the theory does not predict the order of quantum events. However, there is also the further problem that the quantum ensemble can be divided into different matter distributions, each represented by a different eigenfunction basis, and the basis remains indeterministic unless an experimental setup is chosen.

Statistical mechanics is indeterministic out of theoretical necessity because the indeterminism explains the non-linear behaviors, although the connection between indeterminism and non-linearity is often not well understood. If we treat the ensemble quantum mechanically (a linear system), then we cannot explain the non-linearity. The non-linearity of the statistical ensemble contradicts the widespread notions about reduction because a fully reduced system is always linear. Therefore, if nature was fully reducible then the second law of thermodynamics would be false, and biological systems would not display non-linear behaviors. Given that living beings follow thermodynamic principles—e.g., life progresses in a particular direction—this entails that biological systems are not reversible systems, which implies that they are not reducible.

General relativity is indeterministic because it only predicts events or phenomena but not the actual objects (and their properties) underlying the observed phenomena. Relativity is consistent with many possible matter distributions. When this theory is treated as a theory of gravity alone, it is possible to swap objects without making a difference to the observations. Furthermore, if a generally covariant theory of multiple properties is formed (for instance, one that includes both mass and charge) then that theory will permit many matter distributions, which cannot be singled out from the standpoint of the theory's mathematical formalism. This indeterminism will entail that there is room for choice even in a fully deterministic theory—because the theory predicts correctly only from a third-person perspective but not from a first-person viewpoint.

Evolutionary biologists claim that their molecular evolution is based on the ideas of reduction and causal completeness of physical theories. This is a far cry from the indeterminism in physical theories upon which biology is based. Reduction is false because it contradicts non-linearity and the irreversibility of thermodynamic phenomena. Causal completeness is questionable due to probabilities in quantum theory and due to matter distribution problems in relativity. The notion that issues of causal completeness and reduction have been settled in science, and biology is only carrying forward these ideas in the context of living beings, is a mistaken view of science. This also means that notions of reduction and completeness in biology are somewhat premature. When fundamental theories of science revise the formalisms to solve the problems of causal incompleteness and irreducibility, then biology will change as well.

Notions of biological determinism and reduction are based on ideas of science that were developed during Newton's time but have since undergone a radical denial in many physical theories. Unfortunately, Newton's picture of nature endures in the scientific minds and everyone interpreting science tries to relate the newer developments to the concepts about nature that Newton created. This bias in science has of late become a hindrance to a true understanding of nature because often, when the exact significance of the shifts is not apparent, the simplifying assumption is to discard the modifications caused to Newton's view as minor 'perturbations' to an already established rock-solid foundation. We often fail to recognize that quantum theory, general relativity, and thermodynamics question the very foundations of physics—namely the ideas of reduction, linearity, causality, and determinism in science. Newton's theory painted a picture of reality where nature comprised individual, independent objects which could be completely described in a third-person manner using a linear and deterministic theory. Modern physical theories question each of the above three premises. We cannot speak about independent and individual objects due to quantum theory. We cannot completely describe nature from a third-person perspective due to general relativity. We cannot espouse reductionism due to irreversibility in statistical mechanics.

Biologists are often oblivious to these problems. The dominant dogma in biology is that nature is reducible, determined, and objective. The solution to the above problems instead requires the conception of a whole that has an identity aside from the parts. It needs a conception of nature in which the parts are produced by dividing the whole through informational distinctions. Then we need the notion of an object which ties these distinctions into wholes. Distinctions are the sense in which nature has many properties. Identities are that which tie these properties as attributes of individuals.

The realist interpretation of science employs both properties and objects (that possess these properties). But scientific theories don't incorporate the distinction between objects and properties; they only use properties to make predictions and then use objects to give a realist interpretation to these predictions (i.e. that underlying the observations is a reality). The object in question is something that doesn't change even when all its properties are changed, which means that it must be independent of all physical properties although no such object truly exists in science. The properties too are defined independent of other properties, which means that these properties could not represent meanings which are only defined relationally. The solution to indeterminism needs the induction of the idea of an object that is independent of all properties—which means that it is not observable in a third-person manner although it must be observable in a first-person manner. This object would combine observable properties into a whole (and thus form an object), while the properties themselves are defined relationally.

The informational distinctions require a new way of thinking about meaning and matter which I will outline in the next chapter. The meanings themselves have to be produced through choices, which require the existence of something that can produce information. The classical question about which object the properties belong to is now the question of which observer produced the information. While the object ties together properties, the observer ties together meanings. The observer is now the object 'behind' observations. The observer is also the object 'in front' of' the observations. The classical notion of causal interactions between objects is now replaced with a new construct of an observer which communicates its meanings to other observers and thereby causally interacts through meanings. The producer of meanings also 'owns' them, quite like a classical object 'owns' all its possessed properties[6].

As I have already noted above, physical theories are causally incomplete and they require additional causal constructs beyond physical properties, when physics deals with object collections. Information therefore acquires a role when we describe object collections rather than individual objects. This information is generally called 'order' or 'disorder' which can only be associated with object collections and not with individual objects. Information divides and distributes matter, and science is incomplete because it does not adequately describe the division and distribution of matter.

Evolutionary biology also speaks about information, but reduces this information to the physical states of individual objects rather than as properties of object collections, disregarding the fact that notions of 'order' and 'disorder' can only be associated with collections and not with individual objects. Given the problems in this reduction, new notions of information that directly deal with the idea of a collection are needed in science, which in turn entail further revisions to fundamental theories of nature as outlined above. For instance, collections (or sets) have to be described as concepts rather than as physical aggregations of independent parts. The creation of such collections now needs a new role for meanings, which have to be governed by new kinds of natural laws, thereby reintroducing a central role for meaning and altering the nature of laws.

Plato and Aristotle during Greek times thought that material objects were representations of ideas; that there is a table or a chair because the idea of a table or chair precedes the objects. Subsequently, Descartes separated mind from matter in a way that material objects were no longer dependent on ideas. That separation creates many problems even in the explanation of matter. The reintroduction of the mind in science is therefore needed not just to understand how the mind and body interact, but also to explain how the material objects behave when they represent ideas. The problems of quantum theory, statistical mechanics, and general relativity are not directly problems of the mind, although they are problems of information or meaning when encoded in matter. The role that the mind plays in nature thus is not a problem for psychologists or neuroscientists alone, but also for physicists and mathematicians.

The Problem with Natural Selection

Natural selection is defined to be a process by which those organisms that 'fit' better with the environment are selected in preference to those that do not fit as well with the environment. Through natural selection, the evolutionary theory creates order from disorder and it seems mysterious how order could automatically emerge from an entirely natural and physical process. With the discussion in the previous sections, we have now acquired the necessary concepts by which this question can be understood and demystified.

In physical theories, the entire universe evolves according to some mathematical law, but different systems behave somewhat differently when we apply *boundary conditions* to them. A boundary fixes the properties of the system as the edges of the system, which in turn partially or completely fixes the properties of the system within the boundary. A system becomes differentiated from other systems, or its environment, precisely due to its boundary conditions. The key factor in the emergence of an ordered system is therefore not the mathematical laws per se (because these laws are common for the rest of the universe) but the application of a system boundary. By applying different boundary conditions, we can create different physical systems with different properties. A molten metal can, for instance, be shaped into a pipe or a sheet based on boundary conditions; a vibrating membrane will produce different normal modes based on the different boundary conditions applied to it.

This fact about physical systems is pertinent to evolutionary theory because the theory postulates a boundary between an organism and its environment. In the early days of science, boundaries were supposed to isolate a particular system from its environment. This isolation, it was generally acknowledged, is not something physically real, but only a theoretical convenience: we choose to study only a part of the universe, rather than the entire universe. When we isolate a system in this sense, we ignore the effect of the universe outside of it. However, as noted above, this is not the only way in which we view boundaries: we can also see them as boundary conditions that constrain and define the properties of the system. The study of boundary value problems is immensely useful in engineering designs, because we know how to engineer boundaries. The emergence of these boundaries is, however, not intuitive in nature itself. How, for instance, are boundaries automatically produced?

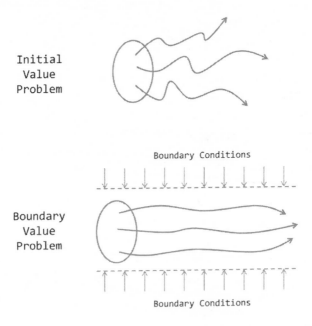

Figure-9 Initial and Boundary Value Problems

The order in evolution is produced as a consequence of converting a physical system from an *initial value problem* to a *boundary value problem.* If we think of the universe as some set of particles which have some initial states governed by mathematical laws, then the system would evolve according to these laws; we don't expect the universe governed by such a law to automatically produce order. The order is however naturally produced when we divide the universe into parts using some boundaries; the matter within the boundary is now constrained by the boundary conditions. For instance, if we have an infinitely long steel wire which is not constrained to move in space, then pushing this wire will possibly cause it to move or deform in many different possible ways. However, if the steel wire's position is fixed at two ends (a boundary condition) then the string will vibrate with a well-defined frequency. The palpable difference between the various possible deformities of an unconstrained wire and the regular oscillations of a constrained wire is the outcome of applying boundary conditions to the wire.

The crux of the mechanism in evolution that creates order is also the application of boundary conditions. The organism without an environment is like a steel wire that hasn't been fastened at its ends; this wire can behave in many possible ways. The organism with an environment is like a steel wire that is fastened at its ends; this wire can only behave in certain limited ways; this limitation is called conformity to environmental conditions by which nature automatically selects. However, natural selection is not itself a physical process. Rather, it is a byproduct of applying boundary conditions. Unlike engineering problems where we conveniently assume boundaries because we design the system in that way, we cannot postulate the natural emergence of boundaries in nature unless we find a mechanism by which nature is being divided in a specific manner.

From set theory in mathematics, it is known that a collection of N objects can be divided into sets in 2^N ways. As the number of objects grows linearly, the total number of possible sets (or boundaries) grows exponentially. These sets constitute boundaries around which the universe is divided. The universe in current estimations contains about 2^{82} particles. This universe of objects can contain an enormously large set of boundaries, subsets, or systems of objects:

$$S = 2^{2^{82}}$$

All of these alternatives are prima facie equally likely, and if we had to pick one of these alternatives in the universe, then we would have to provide S information. Since this information is exponentially larger than the entire universe, to fix one universe, we would require exponentially many universes. It is noteworthy that this information does not exist as additional matter or energy. It is simply the division of the universe by boundaries. Natural selection becomes a viable method of producing order in nature when the universe has been divided into systems using boundaries. However, this also begs the question of how these boundaries are produced.

Natural selection does not provide a mechanism for producing boundaries in nature. It only states that if we can produce boundaries in nature then we would be able to create order naturally. The key question therefore is: Can we produce boundaries naturally? At first sight, this seems quite straightforward. For instance, we can suppose that due to natural forces particles automatically aggregate and the aggregations constitute natural boundaries. But this idea is not as straightforward as it might initially seem. For instance, given an initial set of N particles, there are 2^N ways in which these particles can aggregate into subsystems, each consistent with the idea that a system can be formed by aggregating particles. We earlier encountered this problem and its consequences in quantum theory, statistical mechanics, and general relativity; each of these theories describes laws but the laws underdetermine matter distribution. In other words, physical theories stipulate the ways in which a total amount of energy can be divided into particle states, but these theories do not single out a particular matter distribution. The division of energy into objects represents the act of drawing boundaries within a collection, and these boundaries cannot be predicted.

There is a further deeper problem related to boundaries which arises from the fact that by inserting such boundaries in space-time we break the symmetry and isotropicity of space-time. The space-time symmetry and isotropicity are the reason that energy, momentum, angular momentum, and spin are conserved in physical theories. If boundaries have a fundamental role in nature, then physics has to reconceive the conservation of symmetry properties in nature. This in turn also requires us to reconceive the nature of space-time and there is no current theory that provides a description of nature in which space-time can be inhomogeneous and anisotropic. In short, there is no current physical theory upon which the idea of natural selection can be justified as entirely natural phenomena.

The matter distribution problem discussed in the previous section is identical to the issue of space-time boundary creation, and the emergence of order based on organism-environment boundary in natural selection hinges on the prior definition of boundaries. It is correct to assume that *if* there is an organism-environment boundary, it would also produce order. However, that order is a byproduct of defining the boundary in the first place. No current physical theory provides the mechanism by which boundaries can be created. The idea of natural selection may not be necessarily flawed, although it begs the causal mechanism underlying its assumptions. Furthermore, as we noted above, the addition of a boundary in space-time does not require additional matter or energy or force. In one sense, there is nothing physical about boundaries in nature and in another sense these boundaries produce all the order in nature.

I will later argue that the creation of boundaries in nature presents a mechanism for mind-body interaction because while the boundaries cannot be *perceived*, they can be *conceived*. The boundaries cannot be seen, touched, tasted, heard, or smelt. And yet, their effects can be perceived by the senses. The idea of a space-time boundary therefore represents a pure concept which cannot be measured although its effects can be measured. The mind can now be thought of as the agent that creates, modifies, and dissolves space-time boundaries. Since boundaries exist in space-time, the interaction between mind and matter would be mediated by space-time although space-time itself has to be defined in a new way. Unlike the flat space-time of current physical theories, where all locations and directions are equivalent, the new space-time theory will define space-time in a hierarchical manner in which the universe is divided into smaller and smaller boundaries. The larger boundary is the 'environment' inside which the smaller boundary becomes an 'organism.' The larger boundary determines the boundary conditions for the organism within it, and the organism must be modified to 'fit' into the larger boundary. This need for the organism to adapt within the environment is called natural selection in evolution because the organisms that don't fit would be naturally eliminated.

The key difference between the current notion of natural selection and the idea that selection is produced when the boundary conditions are applied is that while the process of boundary creation (and hence order creation) is natural, it is not physical because the production of the boundary cannot be predicted by any physical theory (as we saw above, all physical theories are indeterministic about matter distributions). In that sense, the boundary exists in a form that is novel for present science because the boundary has to be viewed as a mental construct, which cannot be reduced to any physical construct. A natural but non-physical theory of nature will not just solve the incompleteness of physical theories, but it will also fix the problems of mind-body interaction. It will explain how the mind's effects can be sensed although the mind itself cannot.

Thus, we cannot 'see' the mind, but we can use the mind to produce new effects in matter, which can be seen. These effects will provide the empirical confirmation of a theory of concepts, such that the concepts exist in space-time as boundaries. The boundaries and concepts cannot be perceived, but they have observable effects. The idea of natural selection would now be a consequence of the effects the concepts have on matter and the evolution of species would be governed by the laws of the evolution of ideas in nature.

Evolution's Halting Problem

I have thus far discussed the problems of random mutations and natural selection in the context of physics, and I will now turn to a discussion of the same problems in the context of mathematics and computing theory. The problem in physics is that science is incomplete as far as matter distributions are concerned. These distributions can be thought of as information which organizes matter into individual objects or systems. The problem of information is, however, not unique to physical theories; the problem also appears in mathematics and computing theory. The latter fields provide even more stark examples of why information is different from matter, although information can be encoded within matter. The incorporation of information in matter requires new ways of thinking about mathematical and computational objects, as we will now see.

One such example of the role of information in nature arises from the fact that when matter encodes information and forms a functional system, the system represents a program which must be *finite*. A finite program eventually comes to a halt. In the case of living beings, this implies that all living beings must eventually die. Random mutations, however, can produce infinitely long programs. A program would be infinite if it has a loop from which the program never exits, and thus, for a program to be useful it must be finite.

Perhaps the biggest existential reality of all living species is that they die. Cells are destroyed and recreated. Even molecules like hemoglobin are regenerated and when this process stops the body needs to be fed hemoglobin externally. Advances in genetics hope to extend the duration of our lives, perhaps one day creating a species that never dies. Such a species will have to have the ability to never cease in its ability to regenerate molecules. Or, it might involve the use of molecules that are never functionally incapacitated. But what does the idea of random chance mutations tell us about death? Functionally, a species that never dies is a computer program that never halts. The fact is that all species appear to ultimately die, means that they are computer programs that eventually halt.

From a computing perspective, a program that never halts is bad, because it will never complete the 'purpose' for which it was designed. The purpose, in the case of a living being, does not have to be a higher meaning of life. Quite like a computer program, this purpose can also be materially defined in terms of the transactions the living being performs, which can be stated materially in terms of inputs and outputs of the program. A program that never halts will continue inputting and outputting information without ever terminating. While such programs are never desirable as machines that run forever, they are highly desirable as eternal life for us, if such a concept was ever biologically feasible. If programs were written by randomly mutating computer instructions then at least some of these programs would never halt. Similarly, if random mutations and natural selection can create a variety of living beings, certainly some of them could also be eternal living beings. The question is: Why hasn't evolution created living beings that do live eternally?

It has been shown[7] that, given any arbitrary program, the program will either halt very quickly or never halt. That is, the probability of a program that runs for a very long time and then terminates is vanishingly small, amongst the set of all possible programs. Most of the programs are either very short-lived or eternal. This means that if programs were being produced by randomly mutating computing instructions, then the likelihood of a program that never halts is quite high. On a similar note, if functional systems were being produced through random mutations, then the likelihood of an eternal system would be quite high. But this leads to the question: Why don't we see such functional systems in nature? What in evolution prevents such systems from being automatically created?

Of course, we could argue that if an eternal species indeed existed, and was multiplying in numbers, then eventually there will be too many of that species and they will consume so much of food, water, and other resources that everyone will eventually run out of resources. That would correspond to a computer shutting down because it ran out of power supply. But, until it does actually run out of food, such a species could keep multiplying and growing in numbers. Can there be anything in nature that can somehow 'know' that the members of some species will not die until all the resources are actually finished? Indeed, given that some species has lived a long time, is there some way in nature that can predict that the species will actually live forever and not die after living a long lifespan?

Whether a program is finite is called the Halting Problem in computing theory. It is important to know whether a program will halt before the program is started, because a program that is infinite will never terminate on its own and has to be externally terminated. This raises the question: When should we terminate the program? How do we know that the program is infinite because it may have been tasked with a job that just happens to take a very long time?

Alan Turing proved a famous theorem called the Halting Problem showing the impossibility of predicting if a program will halt (or not halt), if the program is seen simply as a sequence of individual instructions. Note that in any program, the individual instructions are always meaningful although the sequence of these instructions may not be. Programs that loop infinitely are byproducts of the inability to design a machine (or program) that can detect the meaning of the sequence of instructions, given that the individual instructions are indeed meaningful. Since a program's meaning as a whole cannot be determined in advance, there is no physical procedure to find out if a program will eventually terminate. Turing's proof entails that the Halting Problem is not just hard to solve; in fact, it can never be solved using any kind of Turing Machine (mechanical procedure). Even if a program has run a very long time, there is still no way to prove that the program will run forever.

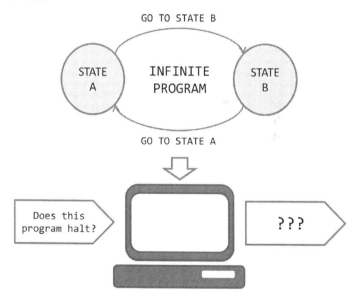

Figure-10 Turing Proved the Halting Problem Unsolvable

This fact implies that there can never be a procedure in nature (if nature is physical things) that can know in advance if a random mutation will create a species that lives eternally. And, therefore, such a species cannot be prevented by natural selection, since natural selection is also a certain kind of physical procedure. Since random mutations imply that infinite programs are imminent, it also follows that if living beings were viewed as computer programs, some of these randomly created programs must also live forever. There is no mechanical procedure that can determine whether a given program is in fact infinite, so if there were an infinite program, there would be no mechanical procedure to actually terminate it. That in turn implies the impossibility of natural selection preventing the existence of infinitely long-lived beings. Therefore, if a natural selection procedure does not exist to destroy eternal beings, and such beings are possible due to random mutations, then they must somehow abound in our ecosystem. Why don't we see them?

An additional problematic fact for natural selection is that at the level of molecules there is no memory of the past. Indeed, a cornerstone of modern science is that causality depends only on the current state, and not on the history of states. When this fact is brought to bear on biology, it implies that there is nothing at the level of atoms or molecules which can ever be aware of history. Since atoms don't have history, they cannot know if an animal has lived for a very long time, and hence eliminate that living being.

One possible mechanism by which a program can prevent infinite execution is if it can discover loops from within itself. This could be achieved, for instance, by using a program counter that increments with every passing tick of time. And the evolutionary system must know how to read this ticking counter. But to even interpret a set of physical bits as a ticking counter, there is need for a programmer who interprets some physical state as the ticking counter meaning. All memory locations in a computer, for instance, are not interpreted in the same way. Memory addresses are accessed by reference, and these references are created by the programmer. If a wrong reference is read, the program is logically flawed and might end up in infinite loops. In short, even for timer detection, there is a need for meanings and references which are not likely to be accurately produced through random mutations. A single false mutation will create an infinite program. Furthermore, even if we suppose that some program could have a timer backdoor for the detection of its age (it should be considered a backdoor because it will eliminate the infinitely lived being), a randomly generated program cannot be expected to always have such backdoors. Such programs will never halt. We can conclude that by default a physical system only involves processes that only look at the current states of atoms and molecules and so it cannot know how long a species has lived. Even if we inject a role for meanings and references into these programs, a randomly produced program might not correctly measure time. Finally, all such random programs are not likely to have timer backdoors. We can conclude that there is no physical procedure that can actually eliminate a species that will live eternally by natural selection.

Infinite programs can only be prevented if nature were a semantic system because then nature could determine if the program indeed has a meaning. A semantically correct program will halt because every meaningful proposition is finite. Accordingly, if nature produces species, and nature understood their semantics, then they will live for a finite time. But, in such a case, we can see that a species is not a random sequence of meaningless bits produced through random chance events. It must rather be produced as an ordered collection of meaningful symbols. In a semantic system, only meaningful programs would be created. Therefore, if nature is a physical system then it will produce infinitely lived beings, which cannot be eliminated by natural selection. If, however, nature is a semantic system, then it will produce finitely lived beings, but there cannot be any role for random mutations. It follows that only one of the ideas of random mutation or natural selection could be true. If nature is physical, then random mutation could be true, but natural selection would be false. If nature is semantic, then natural selection would be true, but random mutations would be false. A theory that postulates both random mutations and natural selection therefore hypothesizes a logical contradiction; it cannot be a consistent theory.

If evolution involves random chance events then it will produce meaningless propositions and programs that never halt which in this case correspond to eternally living species. If, however, evolution involves encoding meanings, then it will only produce meaningful propositions and programs that halt in a finite time, which corresponds to living beings with a finite span of life. This presents us with a simple theoretical and empirical test for evolution theory. The test is that if species are produced by random chance events then eternal species are imminently likely. Furthermore, there is no physical procedure (e.g., natural selection) that could preempt such a random mutation. Therefore, if eternal programs are highly likely and they cannot be preempted by natural means, then they must abound in nature. However, we don't see any such eternal species. On the contrary, we only see living beings that have a very short time-span, at least in comparison to the geological or universal timescales. Why haven't random mutations created eternal species, if these species are highly likely? What in nature prevents these eternal programs? Since we do not find eternal species, the production of these species cannot be based on random chance events.

The key point of inflexion in this argument is what we mean by randomness. I will define it as a sequence of random instructions (which can be encoded using atoms and molecules). Random sequences represent meaningless propositions, not because the individual instructions are meaningless but because collectively they may constitute a program that never halts. The Halting Problem is not about whether an individual instruction is valid. It is rather about whether a collection of instructions is valid. The validity of an instruction depends not just on that instruction but on the other instructions that come before or after it. Given that invalid programs are likely, nature must abound in species that live forever. Clearly, the evidence goes against this. Indeed, the very idea that all species must eventually die because of natural selection logically contradicts the idea that some such species must live eternally.

Alan Turing assumed that the Halting Problem can be solved by a mechanical procedure and then arrived at a logical contradiction. His proof implies that unless we can know program semantics, we cannot know if a program will halt. Since there is no way to know program semantics in current computing theory—because a computer does not hold meanings, but only physical states—the Halting Problem implies that in a physical world nature cannot determine which programs are meaningful (and finite) and which ones are meaningless (and infinite). The notion of random chance events represents a logical contradiction similar to Turing's paradox.

The hypothesis that a random process of creating propositions will only produce finitely lived programs leads to a logical contradiction because many of these random programs must run forever. If programs run forever, then there is no evolution because these species will live forever and cannot be selected by the environment. To avoid this contradiction, biology can postulate either random chance events or natural selection, not both. If we postulate random chance events, then some species will live forever and cannot be selected. If, however, we postulate natural selection, which implies finitely lived programs, then we cannot have random mutations. The extension of Turing's Halting Problem to evolution demonstrates that random chance mutations plus natural selection is logically inconsistent. There can be only one of these, but not both. Of course, evolution will not explain the diversity of species if only one of its two pillars (random chance events and natural selection) were true. And postulating both results in a logical contradiction.

We can conclude that the premises in evolution constitute a logically inconsistent theory. The production of finitely-lived species can only be explained if these species are produced as a process of encoding meaning, which requires the meaning to exist in some form prior to its encoding in matter. If biological forms encode semantic information, and nature understands semantics, then it will only produce meaningful and hence finite programs. That process will not involve randomness and it will explain why only finitely lived beings are created through a natural process in biology.

The Problem of Meaning

The issues of missing causality in quantum theory and the need for finite programs in computing bring the issue of meanings at the center of biology, even from a physical perspective. Of course, the issue of meaning has always been a central problem for evolutionary biology, when biology is tasked to explain the meaning processing capabilities in the brain, and how the brains could have evolved to process meanings when other physical things in the universe are meaningless[8]. Modern biologists widely suppose that at some point in the development of neural complexity, a collection of meaningless things (molecules) becomes meaningful. The fundamental problems involved in this supposition are not widely understood at present.

A fairly common stance in modern science is that individual physical objects don't have meanings, but a large collection of these particles—structured and organized in some unique fashion—give rise to the meaning capacity. This raises the question of what in the large collection of particles corresponds to meaning. What do we mean by structure and organization which gives rise to meanings? The straightforward answer to this question is that all structure and organization in the physical world reduces to some space-time distribution of matter. Physics shows that matter is built up of fundamental sub-atomic particles which are then distributed in space-time to form atoms, molecules, lattices, and other complex biological structures. Now physical theories also claim that the sub-atomic particles themselves are meaningless. They have physical properties but they do not *denote* something other than those physical properties. For instance, the sign '$' is a squiggle. It does not by itself denote the idea of money. Therefore, how can a collection of meaningless things acquire meanings? What *mechanism* underlies the transition from meaningless objects to meaningful symbols?

The idea that random aggregations of physical things can denote meanings works only when the mapping between things and meanings is established outside things. In the everyday world, for instance, the mapping between words and meanings is supposed to exist in the mind. Language provides a set of words or sounds, which can be physically described as frequencies, amplitudes, and wavelengths. If we look at these words physically, they have no meaning. Our minds, however, map these sounds to meanings. This raises two kinds of questions. First, how does the mind map sounds to meanings? Second, if the mind maps meanings to sounds, then in what form do the meanings exist prior to their being mapped? Note how the mapping between words and meanings is outside matter in science, and is claimed to exist in the mind. However, to create this mapping, there must be something called meaning that must exist in the mind. The mind may map things to meanings, but we haven't really explained what meanings are and how they map to things.

This view about the relation between mind and matter leads to problems when the mind is reduced to the brain, because if the mind is indeed the brain then the meanings in the mind must be due to the physical arrangement of molecules and neurons in the brain. We are now back to the original question of how material arrangements denote meanings. If some material arrangement can denote meaning in the brain then a similar material arrangement can also denote meaning in the physical things. What is that arrangement? This may seem like a question about the biological structure of the brain, but it is also a question about the physical basis of meaning. Theories about that physical structure haven't been found. That may not necessarily mean that meanings can't be encoded in matter. It may only mean that we haven't yet found what in matter encodes meanings. An account of matter without the ability to encode meanings is incomplete, and unless the physical basis of such encoding is found, the physical description of matter is itself incomplete.

The problem of meaning, which the mind-body dualist evicted from the physical world into the mind, which the neuroscientist reduces to the brain, which the biologist claims is nothing other than molecules, which the chemist reduces to the structure of atoms, which the physicist claims is mass, charge, energy, and momentum, faces a dead-end today. The biggest mystery as far as living beings are concerned is not their complex physical structure. The biggest mystery is that we are conscious and have the ability for sensation, emotion, intention, and free will. To begin to solve this mystery we can begin to approach the problem by considering only that part of conscious experience which seems most amenable to an objective-empirical study. This is the problem of meaning. Meanings can exist in books, music, and pictures, which are objects. These objects don't seem to manifest emotions, intentions, and free will. So, if we can understand how these objects encode meanings, then we can extend the understanding to how the brain encodes meanings. Over time, the range of meanings can extend to emotions and intentions as well. But even the most rudimentary form of meaning requires a significant shift in how we view nature. Contrary to how we have thought of everything in terms of physical particles, the only way to construct meaning is to start seeing objects as *symbols* of meaning.

Language begins in alphabets or letters which are combined to create words, sentences, paragraphs, chapters, and books. Letters themselves appear to be meaningless, but their combination into words
seems to denote some meaning. This meaning is not fixed; the same
word can have different meanings in different contexts. But by the
time we get to words, there is already some semblance of meaning in
the token. How do meaningless alphabets create meaningful words?
The general view is that the words themselves don't have meanings,
but the mind endows them with meanings. As we saw above, this leads
to the problem of how the brain encodes meaning when the mind is
reduced to the brain. Of course, meanings don't stop with words. The
meaning in a word is fixed by the context of the sentence. The meaning
in a sentence is fixed by the context of a paragraph. The meaning in
a paragraph is often fixed by what preceded and follows in the paragraph. And in case of mystery novels, the meaning of an utterance at
the beginning of a book is decoded only at the end when the entire
picture becomes clear.

These facts about meanings imply a contextual view in which
meanings exist in the interrelations to other symbols. Science, however, describes objects independent of contexts. An object's physical
properties—in classical physics—are not changed by the presence of
other physical objects. In quantum theory, the properties in objects are
changed by the presence of other objects in an ensemble, but quantum
theory still describes objects in terms of physical properties of independent objects, not in terms of meanings. To address the problem of
meaning encoding, we are required to change the manner in which
we describe material objects. This change will pertain not just to the
brain but to any physical object whatsoever. That in turn requires a
shift in physical theories. Unless that shift is made, attempts in biology to reduce meanings to matter are incomplete. We are essentially
measuring the height and weight of a book, not truly decoding the
meanings in that book. Similarly, attempts in neuroscience to reduce
the mind to the brain are also incomplete, because the brain too is
described as the physical properties of atoms and molecules and not
in terms of their meaning encoding.

An evolutionary theory that tries to explain the evolution of the
brain without first explaining how matter can encode meanings is

fundamentally incomplete because the basic property of the brain is to encode meanings. Unless we understand how the brain encodes meanings, speaking about the evolution of the brain is like talking about the creation of a book without understanding the meanings that the book encodes. The order of squiggles in the book is not random. Rather, that order represents a new kind of property which cannot be understood unless the words themselves are seen as symbols of meaning. If the squiggles in the book have no meaning, then we can measure squiggle probabilities but we cannot predict the order in which the squiggles will appear[9]. Therefore, the prediction of matter distribution in an ensemble—when the distribution denotes meanings—will remain statistical and incomplete. The reinterpretation of matter in terms of meanings is therefore not just another way of looking at reality without an empirical consequence. Rather, the semantic viewpoint can help overcome statistics.

This shift in physical theories also entails a fundamental shift in how we understand biology and the brain. The meaning processing capability in the brain is not just a byproduct or epiphenomenon of aggregating meaningless particles, but a fundamental property of the sub-atomic particles themselves. This shift in thinking also dramatically changes our outlook towards evolution. When matter encodes meanings, the laws that govern its evolution are also different. In particular, what we currently call randomness will be a consequence of the fact that current science attempts to describe semantic phenomena in terms of physical properties and this description is incomplete. Currently, we attribute the incompleteness in our physical theory to randomness in nature when the real problem is that we are trying to describe semantic phenomena in terms of physical properties. If nature is semantic, then what we call randomness is a deficiency of our theory not of nature. To fix the theory we need to shift our approach to nature from describing things to describing symbols of meanings. The evolution of symbols is governed by different laws which current theories do not capture.

The idea that nature is meaningless particles which aggregate randomly to form meaning processing brains involves a theoretical transition from meaninglessness to meaningfulness which can never be made in today's science because a physical object can never represent or refer to another object. All objects are things-in-themselves and never things-about-things. The simplest meaning processing capability in the brain requires the existence of *names* and *concepts.* Names refer to other things and concepts represent other things. Both names and concepts can be encoded in material symbols, but such symbols have representational and referential properties which are impossible in a physical understanding of nature. No amount of physical complexity can convert an object to a name or a concept. Rather, we have to revise our notion of matter from meaningless things to symbols that encode names and concepts. In this theory, there will be no room for random occurrences because randomness is an outcome of a failed attempt to reduce meanings to physical properties (since the reduction is incomplete we tend to believe that some occurrences in nature are random).

A semantic theory will eliminate randomness and all changes would be lawful. The evolution in nature will also be law governed and not random, although these laws will be different from the laws that describe the evolution by physical properties. The problem of meaning in the brain thus changes our view of matter. This view fixes the problem of the incompleteness—and hence the randomness—of predictions in nature. However, once we discard randomness then we must also discard the theory of evolution itself.

Incompleteness in Mathematics

Perhaps the most convincing argument about the role of meaning comes from logical paradoxes in mathematics. There are many paradoxes in current mathematics, which I have separately surveyed[10]. I have tried to show in my previous analysis that logical paradoxes arise through what philosophers call category mistakes. Ordinary language comprises many categories. For instance, the distinction between things, names, and concepts is part of all ordinary languages. By this distinction, a word sometimes is a thing, sometimes a name and sometimes is a concept. Take the word 'nobody' for instance and consider the following three statements that use it.

- Nobody has six letters

- Nobody is perfect

- I am nobody

In the first statement, the word nobody is a thing, a sound, or a physical aggregation of six letters. In the second statement, the word nobody represents a name that calls out or refers to people. In the third statement, the word nobody represents the concept of insignificance. Ordinary language is replete with such uses of words. If we were to use these words interchangeably, we can come up with interesting conclusions such as the following two shown below.

- Nobody is perfect. I am nobody. Therefore, I am perfect.

- Nobody is perfect. Nobody has six letters. Therefore, perfect has six letters.

In the first case, we have arrived at a false conclusion but it is not logically incorrect. In the second case, we have concluded something that is logically incorrect, because the word 'perfect' has seven letters and not six. We don't need to be linguists to understand the source of these paradoxes. The problem here is that the word 'nobody' can denote things, concepts, and names. To use language consistently, the user must respect these categorical distinctions; when that distinction is not honored, paradoxes inevitably arise.

In mathematics, too, categorical confusions appear when numbers are interchangeably used as objects, names, and concepts. The number '5' for instance can be treated as a *thing* or as a token. It can also be treated as a *name* when we have ordered objects in a collection and labeled them by numbers; for instance, we can call someone 'Employee #5.' The same number also represents the concept of fiveness, which is present in all collections with five objects. Mathematics has the ability to talk about each of these meanings independently, but not the ability to talk about them at once. In separate contexts (or theories) mathematics can formulate and describe ideas about numbers as things, names, and concepts. But when the same theory employs all these categories at once, paradoxes are possible. Well-known examples of such paradoxes include Gödel's Incompleteness Theorem, Turing's Halting Problem, Burali-Forti Paradox, Russell's Paradox, and Tarski's Undefinability.

Mathematical paradoxes point to the fact that mathematics lags behind ordinary language in its ability to deal with linguistic categories. To deal with these categories, it should be possible in mathematics to treat the same number as a thing, name, concept, program, algorithm, problem, and perhaps other things. We might say that a number is a physical thing or a token which has several different kinds of meanings. These meanings are sometimes called *models* of a symbol and they represent different interpretations of the same symbol. In the formal study of models in mathematics, a *theory* is a collection of sentences whose truth is decided by the syntax of that sentence. A *model* on the other hand is a domain of objects that satisfies the statements of the theory. For instance, we can formulate a theory of numbers which describes the outcome of addition and multiplication operations. Such a theory will make claims such as:

If A and B are numbers then A + B is a number

If A and B are numbers then A * B is a number

The claims of this theory are satisfied for different kinds of domains such as natural numbers, integers, rational numbers, and irrational numbers. These domains are called the *models* of the theory. If we had to explain the theory, we can concretely demonstrate it first by showing the addition and multiplication of integers. Then we can show that it works even for rational and irrational numbers. These different models are various ways to *understand* the theory, but the theory is different from that understanding. Inherent in mathematics is therefore the separation between the statements of a theory and the *meanings* that can be attached to the statements.

Generally, we assume that the various models or interpretations of a theory are mutually exclusive. That this, different interpretations of a theory are true as long as we keep these interpretations separate from each other. This assumption is false for ordinary language where various categories applied to words are present in language *simultaneously*. Thus, for example, we don't use different contexts for things, concepts, and names. Rather, we talk about things, concepts, and names alternately within the same context. This leads to paradoxes because claims made about a thing can be applied to its name which is then interpreted as a claim about a concept, resulting in a contradiction. For instance, Gödel framed a sentence "P is False." This statement was now named "P" and the name was interpreted as the idea that P is False. Now, it appears that statement P claims that P is false, which is a logical contradiction.

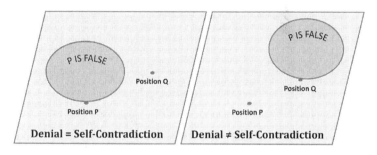

Figure-11 The Conflict Between Meaning and Naming

This problem would not arise if one of two things were possible. First, if we could maintain the distinction between names, things, and concepts, then the contradictions would not arise. Second, if we could prevent "P is False" being named as "P," then even the dissolution of categorical differences would not lead to contradictions. Neither option is allowed in current mathematics. The inability to apply the first option results from using mathematics like ordinary language but not having its distinctions. Like ordinary language permits the use of different categorical distinctions simultaneously, but somehow maintains the distinction between them, mathematics too can be used with categorical differences but since the distinctions cannot be maintained in mathematics the result is contradictions. The second option is forbidden because mathematics measures meanings using the syntax of statements rather than their semantics. In this approach[11], there is no way to know that the meaning of the squiggles "P is False" is that the statement P is false. When a statement is syntactically analyzed, its meaning is not understood. For instance, in another language, the word 'False' could mean 'True' and syntactical analysis would still not detect it.

This brief survey of mathematical incompleteness is by no means exhaustive or even rigorous. But, hopefully, it shows why it is *logically* impossible to construct *representational* and *referential* theories. Names are referential and concepts are representational. A representation *denotes* meanings through symbols, but does not reduce meaning to those symbols. The impossibility of achieving this within the current mathematical theories is given by the fact that any representational theory of descriptive meanings will need to incorporate at least the categorical distinction between things, concepts, and names to avoid contradictions. The representation of prescriptive meanings (i.e. programs) similarly requires a categorical distinction between programs, algorithms, and problems[12]. There are other categories in ordinary language that go beyond description and prescription and these too will need to be eventually invoked in a representational theory. However, there is no way in current mathematics to keep these categories distinct, because the categories denote the meanings of a symbol which is different from the symbol and mathematics cannot put all these categories in the same theory as the symbols themselves. The introduction of categorical distinctions within a symbolic theory creates logical contradictions as Gödel proved through his Incompleteness Theorem.

To solve these contradictions, mathematics needs a theory of symbols in which different types of meaning of the symbol can co-exist with the symbol. That is, the symbol and its meaning are not identical (because then the categorical distinction between things and meanings will not exist) and the symbol and its meanings are not separate (because then we will not be able to speak about the meaning of the symbol in the same theory we speak of symbols).

Mathematical paradoxes illustrate that mathematics needs to hold categorical distinctions (i.e. differences between various classes of meanings) but it is not capable of holding meanings as long as it describes symbols physically. We can also conclude that any theory that describes reality in terms of physical tokens will be either inconsistent or incomplete if those tokens are also required to denote meanings. If the theory is consistent, then it will be incomplete because it cannot incorporate categorical differences. However, if the categorical differences are incorporated in the physical model, then the theory will become inconsistent. This is a far bigger indictment of physical theories than what biologists acknowledge. The limitations are not just about number theory. They are problems for any scientific theory that describes reality as physical properties.

Most evolutionists don't seem to give much serious thought to this problem. They don't seem to realize that it is *logically impossible* to formulate a theory of meanings using physical properties alone. No concept or name used to denote objects can *logically* exist in a universe comprised only of physical properties. This is not a question about how much complexity exists in the brain and how that complexity has evolved over millions of years. Fundamentally, it is logically impossible to formulate a description of meanings from a theory of things because names and concepts are different categories, and they cannot co-exist if nature is just material objects.

Philosophies that reduce mind to the brain commit the categorical mistake of equating names and concepts with things. This reduction begs the question of how a thing becomes a concept or a name. It is a fallback to the original problem of encoding meaning in matter. If, however, we ignore that problem and insist that things can encode names and meanings, then this leads to logical contradictions. Without a clear explanation of how things encode names and meanings, the theory of the brain is incomplete. If we assert that the theory of things is the theory of names and meanings, then the theory is inconsistent. Current biology can either be incomplete or inconsistent. Without an explanation of the mind, biology is incomplete. If, however, biologists reduce the mind to matter then biology will be inconsistent. It is noteworthy that the problem will not be solved by increasing the complexity of the theory or by studying deeper structures in the brain. As Gödel's Incompleteness shows, *any* theory that commits these category mistakes will be inconsistent or incomplete. The question in biology should therefore be: How can we solve the category mistakes rather than how can we delve deeper into the structural questions of brain architecture?

Functionalism and Second-Order Logic

When biology uses the idea of functions, there are two kinds of languages involved. The first language speaks about objects while the second speaks about the functions performed by these objects. The functions are effects of the objects, and they can therefore be attributed to the object's properties; e.g., in physics, we might say that the object's effects are due to its energy, momentum, temperature, mass, charge, etc. In general, an object can have many effects, and to explain them we postulate multiple properties. These effects can also be created by other objects with similar or different properties; thus, an object can have many properties and many objects can have the same properties and an object can perform many distinct functions and a function can be performed by many different objects[13]. This fact requires a clear distinction between an object and its properties: objects cannot be identical to their properties. The irreducibility of objects and properties has some peculiar implications for biology: it means that biology too needs to use both of them.

OBJECT: PROPERTY: OBJECT: PROPERTY:
IS-A CAR   *HAS-A* STEERING   *IS-A* STEERING *HAS-A* SPOKE

Figure-12 Object and Property Distinction

Even in everyday language, the distinction between objects and properties appears in the divide between *is* and *has*; objects are the *is* and their properties are the *has*. For instance, we might say that this is a human, and she has a hand; or, that this is a hand and it has fingers; or, that this is a finger and it has a nail, etc. Similarly, we might say that this is a car, and it has a steering, or this is a steering and it has a spoke. In this distinction between objects and properties, the objects are the whole and their properties are the parts of that object. Of course, the distinction between object and property is not fixed permanently and we can divide an object into smaller and smaller parts, thereby creating more and more contingent objects in a hierarchical manner; any given node (except the root and leaf in a tree) is a property in relation to the higher level node and a property in relation to the lower level node. The root node is always an object and the leaf node is always a property. Except for the leaves and the root, every other node is both an object and a property.

The distinction between properties and objects becomes necessary because the same thing can be viewed as a property or as an object, depending on whether it is related to a higher or lower node in a semantic tree. Thus, context determines whether something is an object or a property; in different contexts, the same thing would be alternately an object and a property, and without the distinction between objects and properties we can create contradictions.

The distinction between objects and properties exists in evolution as well. Here, the notion of random mutation presumes that something is mutating—e.g., there is a DNA *molecule* which is undergoing change. We do not claim that one DNA molecule disappears while another one reappears; we rather say that a molecule underwent a mutation, implying that the DNA molecule in some sense persisted even though the molecule's properties changed and this claim requires the distinction between objects and properties. Similarly, natural selection implies that the object as a whole is being selected rather than its properties. There is nothing in the properties that can be selected—after all, all properties (mass, charge, energy, momentum, etc.) are just physical states and they cannot be selected. What can be selected is a particular *combination* of these properties, and this combination represents an object. While the mechanism by which these properties are combined into an object is not very clear in science today—and what an object *is* cannot therefore be defined—the assumption that there are indeed objects that combine properties is essential to the vocabulary of science.

In fact, the need for object-property distinction is deeper in evolution than in other areas of science. The need is caused by the fact that random mutations occur in the object properties but the natural selection occurs on the objects (or the living being as a whole). We can observe the changes to individual properties of objects but these changes are not natural selection. The state changes in individual particles are no different than the motion of objects in an ensemble. To speak of natural selection we must shift the focus from the parts of the ensemble to the ensemble as a whole because only the living being as a whole is selected, regardless of how many parts or properties it might have. Natural selection is therefore a concept alien to physical theories because it requires us to postulate the existence of *wholes* beyond the individual *parts*. While the parts can individually mutate, only the whole can be naturally selected.

Therefore, if natural selection has to be a real physical process, then the whole too must be a real scientific concept. Unfortunately, no physical theory permits the reality of ensembles or collections. Only the individual particles are believed to be real, although even the reality of the particle cannot be proven; we can only speak about the reality of the individual properties, but how these properties are combined to constitute an object is beyond a physical theory, unless the theory acknowledges the reality of ensembles or collections. If the whole system is regarded as the real object, then its parts constitute its properties. The object in question is not a point location at which several properties—such as energy, mass, momentum, angular momentum, charge, spin, etc.—are concentrated. Rather, the object in question must be the collection of all these properties, and each of the properties can themselves be described as distinct objects, when they are divided into smaller and smaller parts.

The object-property distinction becomes problematic for two main reasons. First, it makes the object necessarily a more abstract idea than the properties of the object; for instance, the car is more abstract than the steering in it, and the steering is more abstract then the spokes in it. Second, when we distinguish between objects and properties without inducting contextuality—i.e. without knowing when something is an object versus a property—the theory that incorporates this distinction becomes logically inconsistent.

To speak about both objects and their properties we require *second-order logic*, which distinguishes between objects and their properties as opposed to first-order logic which uses only objects or properties. In language, we refer to objects by their names and to the properties of these objects using concepts. The use of second-order logic implies that language must have the ability to distinguish between the *names* that call out objects and the *concepts* by which the presence of those objects is observationally detected.

It is a well-known corollary of Gödel's Incompleteness Theorem that there is no consistent and complete mathematical theory that employs second-order logic. The reason is that a word can alternately represent names and meanings, and without a mechanism for distinguishing when a word refers to a name or to a meaning, the language that uses both names and meanings is inconsistent.

Gödel's Incompleteness has some dramatic consequences for biology, which so far have gone unnoticed. The implication is that if we use only objects, then the methods of first-order logic are employed, and Gödel showed this logic to be consistent and complete. Physical theories also require the distinction between objects and properties, where objects are names and their properties are concepts. When the distinction between things, names and concepts needs to be inducted, then second-order logic is needed and Gödel's proof shows that there is no consistent and complete theory of second-order logic. If we use only objects or properties, then the theory is consistent but incomplete, and this incompleteness manifests in indeterminism, irreversibility and incomputability. If, however, we use both objects and properties, then the theory is inconsistent.

In the context of biology this means that if we describe a living body only as objects or properties, our description will be logically consistent but the description would be incomplete. If, however, we use both objects and properties then the theory would be inconsistent. Specifically, in the case of evolution, random mutations can occur on properties but natural selection must occur on objects, which in turn necessitates the use of second-order logic, although the theory can never be consistent and complete. If we use only random mutations or only natural selection, the theory can be consistent but incomplete. If we use random mutations and natural selection then the theory would be inconsistent. In other words, evolution is a flawed theory when the living being is treated as a physical system because when the world is treated physically then there cannot be a mathematically complete and consistent theory of evolution. This problem is logical and conceptual and has nothing to do with the living being itself or the nature of life; the problem arises simply because our language in science presupposes that there are both objects and properties but there is no consistent and complete description of the world when both of these need to be employed.

A complete and consistent description can, however, exist if nature is treated semantically although this treatment necessitates the idea that collections are real properties in nature and not merely epiphenomena of the parts. The collections would not denote more abstract concepts relative to the parts. For example, the hand is more abstract relative to the nails and the body is more abstract relative to the hand. To employ the semantic treatment, we have to construct a semantic hierarchy in which reality is accorded to ever more abstract entities all the way up to the root of the tree. The higher nodes in the tree cannot be perceived by the senses, but they must be real if our science has to be consistent and complete. New methods of perception must now be developed to observe the ever deeper forms of reality—deeper here indicates more abstract.

Biology already employs a semantic description of the living being when it treats the parts of the living beings *functionally*. For instance, we say that a certain molecule is an enzyme because it plays a certain role within a collection of parts, and this role would be different if the whole system were to be changed. In other words, the physical *properties* of the molecule would remain unchanged (if we viewed these molecules as classical objects) although their *effects* would dramatically change. Chemists describe the molecule as physical properties while the biologist describes it as effects. It is generally assumed that the effects are produced from the properties, and the properties are independent of the other objects. This assumption is false—the effects are clearly different in different collections and to account for these differences we must also postulate that the properties of objects must be different, too. In other words, we cannot suppose that an object's properties are not independent of the other properties in the collection; these properties can no longer be treated physically; they must be treated semantically.

Quantum theory is useful in modeling such phenomena, because the properties of a quantum depend on the ensemble as a whole. In that specific sense, quantum theory breaks away from classical physics, although the problems of indeterminism in quantum theory have resisted a clear solution to this problem. The quantum should not be viewed as a thing that exists independent of the other things; the quantum must rather be viewed as a function or role within a collection. The role or function is necessarily defined only in the relation to the collection, and in that sense the collection is real over and beyond the parts in that collection. In fact, the collection must be logically prior to the parts within the theory. In biology this means that the living being is prior to the body parts.

Biologists frequently equate the functional language with the physical language—they argue that an enzyme is nothing but the molecule which is nothing but the properties of the constituent atoms, which are nothing but physical entities that exist independent of the other entities. The problems of functional language in biology are rather deep—they don't begin in biology but in a fundamental misunderstanding about the nature of wholes and parts even in physics. The parts should be treated not as independent particles but as functions and roles that are defined within a whole, which is then defined in relation to a larger whole, each of which is conceptually more abstract than the previous contingent entities.

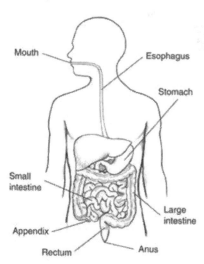

Figure-13 Functions Defined in Relation to the Body

The problem of the use of functional language in biology can therefore be summarized as the following three-step argument:

- Functions or properties are necessary to evolution because random mutations occur on these functions or properties.

- Without an object description, there cannot be natural selection, because the selection only operates on the whole object.

- The necessity for both objects and functions requires the use of second-order logic and Gödel's incompleteness shows that there is no consistent and complete theory for second-order logic.

The idea of natural selection and random mutation therefore cannot be used to construct a logically consistent and complete theory of evolution. If and when a second-order logical mathematics can be constructed, it can also be used to speak about evolution. As we discussed earlier, such a theory requires the induction of semantics into mathematics such that mathematics can distinguish between names and meanings as distinct categories. The separation of names and meanings will enable aboutness in biology since an object that denotes a name can refer to an object that represents a meaning. The mechanism for natural selection will now be intentional relations between objects. A system of intentional relations will constitute an ecosystem, in which parts of the ecosystem cannot change independently without changing the other parts in the ecosystem. Even in the intentional system, therefore, natural selection will *prevent* the emergence of new species rather than create new species, unless the ecosystem as a whole evolves to a new state. The macro evolution, however, would not use random mutation or natural selection.

We saw in earlier sections that random mutations are symptoms of current physical incompleteness and they arise when a semantic system is described physically. We also saw that randomness will not exist in a semantic theory. We have now seen that the idea of natural selection cannot be grounded in any known physical theory of objects because functions cannot be reduced to objects (although objects can be reduced to their functions contextually). Moreover, if objects and functions have to co-exist in biology—just as current evolutionary theory postulates—then there cannot be a logically consistent and complete theory due to Gödel's incompleteness. These problems can be solved in a semantic view, allowing both objects and functions, although natural selection will be a mechanism that reverses arbitrary mutations: the mutants that don't fit into the macro evolution will be eliminated by selection.

Natural selection involves a functional language, which can only be used in a semantic theory. The semantic theory will forbid random mutations since the theoretical incompleteness of current physical theories would be solved. A mutation is now a change in the meaning denoted by an object, and that change is lawfully governed. The change must also be consistent with the larger ecosystem changes in meaning, if that change has to survive. Now, natural selection ensures consistency between the parts and the whole and the biological evolution becomes the evolution of the wholes.

The Genesis of the Functional Problem

It might seem surprising to some readers that there is no consistent and complete theory that incorporates both objects and properties. Isn't all of science only about objects and their properties? How can science exist if there is no consistent and complete theory that incorporates both objects and their properties? The answer to this quandary is that current physical theories only speak about properties and never about objects. The notion of an object is added *post hoc* by the physicist to give a realist interpretation to the physical theory, but the notion of an object never enters a physical theory.

Take gravitational theory for example. It speaks about the attraction and repulsion between masses. The mathematics of gravitation involves the measurement of mass and acceleration. However, physics textbooks also speak about a *particle* which *possesses* a mass, but, in actuality, this particle never enters the theory of gravitation. For this reason, physical theories are only theories about physical properties, because objects are never needed inside the theory. Objects are only needed to give a realist interpretation to science, namely the idea that science is about the nature of reality. Physicists, for instance, will speak about reality as if it were really comprised of particles once the gravitational theory has been empirically confirmed although the theory only dealt with the attraction between masses and never in the idea of the particle itself.

Most philosophers of science recognize the problem of realism in science, and they question whether there really is a particle or field underlying the measured properties by which their presence is detected. When physicists are asked the same question, they will generally throw up their hands and say: It doesn't matter whether there is a particle or a wave, as long as the theory predicts correctly! The debates about realism have not had an impact on the practice of science, partly because the problem itself has not been framed correctly. The problem is that if we truly have objects and properties then physics would need to use second-order logic, which would entail that all natural theories are inconsistent or incomplete. Of course, all natural theories are already incomplete although we haven't yet traced this incompleteness to a fundamental *logical* problem; we still think that there is something missing from our understanding of nature, although the missing piece has nothing to do with our fundamental assumptions about the nature of reality.

If there is indeed a particle apart from its properties, then that particle should remain immutable or unchanged even if all its properties were to change. That in turn would imply that we can make all the properties of the particle zero and the particle would still exist even though we can never know or measure its existence. Now we have added a metaphysical assumption to science that supposes things to exist even though we can never detect their presence.

It is a fortuitous fact of science that it began by postulating objects, reduced objects to properties, and studied only the phenomena where such types of reductions could be achieved. As we saw above, this reduction is possible when an object does not have contextual properties. Since contextual properties arise only in object collections, it is possible to ignore them while studying objects in a manner that the effects of collections can be ignored. Newton's physics is such a theory; it describes independent objects. When, however, we collect such independent objects, new properties arise in the collection, but classical physics is incapable of dealing with them. We earlier discussed how irreversible thermal phenomena cannot be reduced to the interaction between classical particles. Indeed, Maxwell and Boltzmann formulated theories of ensembles of such particles showing that the collection is an uncertain state even though we suppose that the particles are in a definite state.

This conflict between parts and wholes has never been resolved; it is now appropriate to suppose that both the whole and part are in an uncertain state and the uncertainty pertains to the amount of information that can be added to the system before the system runs out of its information storing capacity. That premise, however, would entail that the system is actually information and not things, and the state of an information storing system is uncertain not because there is lack of information but because the information is *abstract*. Just as the idea of a 'car' incompletely specifies the nuances of a particular car even though the car is a definite concept, similarly, a collection is a definite object even though classically we think its state is uncertain. To understand this object we must view it as an abstract concept which can be refined by adding more details. The details—when added to the abstract—acquire their properties in relation to the abstract. The details are not independent of the whole and they become functions or roles in the whole.

Biologists, too, speak about the different functions in the living body, which are contextual properties of chemicals. If we go about examining the living body we will find chemicals, and it is therefore compelling to think that the body is nothing but a bag of chemicals. However, the key question is whether a molecule has contextual properties or just possessed properties. If a molecule has contextual properties then its state—e.g., the composition of the molecule and its specific folding pattern—is itself a contextual and not a possessed property. If the other molecules were removed from the ensemble, the chemical composition and folding patterns would change. That tells us that a molecule's structure cannot be determined without reference to the other molecules in a collection.

Chemists and physicists will generally argue that the contextual properties in molecules are due to the physical forces that each object exerts on the other objects. These forces, they will argue, depend on each object's possessed properties, which are independent of the other objects. This was indeed a picture of reality true in classical physics, but it is false in quantum theory where how many and which objects exist *a priori* cannot be asserted. The same ensemble of energy can be divided into many different sets of objects, so it is impossible to assert that there is indeed a set of particles with possessed properties which then exert forces on each other to influence their states. In the quantum picture of reality, we cannot suppose that there are individual molecules which exist independently of the other molecules and then aggregate into collections through random combinations, and that the energy of the interaction is due to the properties of individual objects. Rather, there is an ensemble of energy which is divided into individual molecules through some mechanism that current physics does not know. Like a fixed amount of ink can be used to write different books, the same energy can be used to produce many different molecular structures. The redistribution of energy into molecules represents information which appears as the choice of eigenfunction basis in quantum theory. It is the information that produces molecules and their structure.

When the parts are logically prior and the whole is produced from their aggregation, the functional role played by a part in the whole is an accident of the aggregation. When the whole is logically prior and the parts are created by the addition of information to the whole, then the functional role is produced by information. In the physicalist view of evolution, the parts are real and the information is epiphenomenal; that physicalist view will entail that we cannot use natural selection because this process must operate on the whole collection and the collection is an epiphenomenon of the parts. The evolution must now occur purely due to random mutations, and any evolutionary scenario without natural selection would be causally incomplete. If, however, both random mutation and natural selection are used, then second-order logic is necessary, which requires us to distinguish between objects and their functions, which in turn requires integration between physics and semantics. This integration requires not just changes to mathematical theories (to incorporate the distinction between things, names, and concepts) but also a revision to the scientific outlook of nature (in which nature is viewed as a collection of meaningful symbols rather than a collection of meaningless objects). The mathematical, computational, physical, and biological problems therefore form a continuum of issues that indicate the need for a shift in our thinking without which all theories will remain inconsistent or incomplete.

However, once the shift has occurred, the current theory of molecular evolution could not be true. It is not that there won't be a theory of evolution; rather, the theory will deal with the evolution of meanings rather than with the evolution of things. Such a theory will deal with closed space and cyclic time rather than with open space and linear time. For biology, such a theory will entail the evolution of ecosystems rather than the evolution of the parts.

Quantum Theory and Functions

The reductionist approach to functional wholes can be deconstructed into three parts—(a) that the parts act independently of the whole, (b) that the parts are logically prior to the whole, and (c) that the whole does not have a reality aside from the parts that constitute them. These three ideas about matter originated in classical physics, which modeled reality as a collection of independent particles whose state did not depend on the state of other particles. Each of the three ideas is false in atomic theory. To understand how quantum theory revises the nature of whole-part relationships in nature the following key facts about the theory must be understood.

First, we must understand that quantum theory describes reality as being comprised of orthogonal eigenfunctions, each of which represents a unique particle. The uniqueness of the particle is that it is logically orthogonal to other particles. All particles in an ensemble are entangled with each other and changes to a single particle's state will potentially change the state of every other particle. This implies that the particles are not independent as they were in classical physics; rather, the states of particles are defined collectively.

Second, we need to recognize that quantum theory supports many different eigenfunction bases, which are equivalent descriptions of a quantum ensemble. Each basis denotes a different *distribution* of matter inside the ensemble and a basis can be chosen by how the ensemble is observed. For instance, in the slit-experiment, different eigenfunction bases must be used to describe outcomes of experiments that utilize different number of slits. By changing the observational apparatus, we change the result of the observation. Now, we cannot assert that there is a definite set of particle states that the observation reveals because by changing the number of slits a different set of particle states will be observed. Given the observer-dependent nature of the measurement outcome, we cannot claim that there is some reality that exists independent of observation. Of course, there is an ensemble of total energy that is real. However, this energy is *represented* differently in each observation, although in some sense it is the *same* ensemble being measured. We cannot assert that the observed parts are real even prior to measurement, although we can still assert the reality of the whole ensemble. Measurement outcomes represent the whole ensemble in terms of their parts, and each measurement procedure divides the whole into a different set of eigenfunction parts. This fact implies that the parts are no longer *a priori* real as they were in classical physics. Rather, the parts are created at the point of observation.

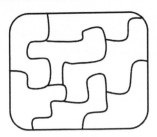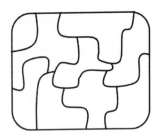

Figure-14 The Quantum Jigsaw Puzzle

Classical physical ideas about part and whole are overturned in quantum theory. In classical physics, the parts were real and the whole was reducible to the parts. In quantum theory, the whole is real and the parts are constructed by dividing the whole. Think of a picture that can be cut into parts in many ways to create different jigsaw puzzles. Before the parts of the puzzle can exist, the whole must exist. The quantum parts are like parts of a jigsaw puzzle and the same picture can create many puzzles. Quantum theory is indeterministic about how a whole is represented into parts, since the same whole can be divided into parts in several distinct ways.

This inversion of the whole-part relation in physics has some key implications for evolutionary theory. In particular, the idea that nature created the parts which then randomly combined into wholes is now inconsistent with quantum theory. We must rather insist that the wholes are more real and they were created ahead of the parts. The parts were then produced by dividing the whole in a way that quantum theory does not explain, because these ways of dividing the whole remain 'choices' of the eigenfunction basis. Therefore, the division of the whole into parts needs a new theory.

As I earlier described, such a theory is possible if nature is seen informationally. The whole and the part now are abstract and detailed information, respectively: a whole is divided into parts when information is added to an abstract object. In quantum theory, the 'choice' of eigenfunction basis represents missing information, by which an abstract object is divided into contingent objects (the whole is abstract and its parts are contingent). Since current physics only deals in contingent objects, it cannot describe abstract information. The abstract information is physical, too, but it can only be represented in science if we describe nature as symbols instead of things. The 'choice' of eigenfunction basis, therefore, does not denote a transcendent, non-material consciousness, but an aspect of matter that cannot be captured by the current physical theories.

It has often been recognized in biology that the functional description of a living being is different from its material description. For instance, it is not necessary that a given type of function—e.g., vision or digestion—is always achieved using a particular set of material ingredients. It is possible that a different set of ingredients could also produce the same function. It is also possible that the same material ingredient could play different functional roles in different systems. Functional and material descriptions are therefore not identical because each function can have many different material realizations and each material object can play different functional roles. But biologists tend to ignore this distinction between functional and material descriptions, attributing reality to the material ingredient and not to the function. Note that the functional description arises when the whole is logically prior and is divided to create functions, while the material description arises when the parts are logically prior and are combined to create wholes. The functional and material descriptions use different assumptions; they involve either semantic or physical presuppositions about nature.

A whole can be divided into parts in many distinct ways, and a part can play many different roles in different wholes. The dichotomy of the part-whole relation is therefore equivalent to the tension between functional and materialist descriptions of a living being.

Quantum theory suggests that the whole is logically prior to the material parts. That is, there is the notion of a living being before there is a notion of eyes, nose, digestion, respiration, or circulation. This might sound strange if we think of the nose or eyes as chemical mechanisms that combine to create the living being, but it is not strange if we think of eyes or nose as information that is produced by adding details to the whole being. The type of the whole and the nature of the information added to that whole will produce different kinds of eyes or noses, but they will be eyes and noses because they are defined in relation to the whole being. We cannot, therefore, suppose that a random collection of particles forms an eye or nose, which then randomly combine to produce the living being. Rather, we must suppose that there is a living being which is first divided into eyes and nose, which are then gradually refined to create the body. The concept of senses—eyes and nose—follows the concept of the living being, and the actual realization of the eye and nose (i.e. the body of the living being) follows the creation of the senses. A new kind of natural mechanism must be postulated in science to explain how a detailed object is produced from an abstract object.

Quantum theory tells us that the whole is logically more real than the parts, but it does not tell us how to interpret these wholes. When that problem is fixed in atomic physics it will have rippling effects in biology as well. Specifically, it would then be possible to speak about how a particular type of eye or nose is produced gradually by refining the idea about a living being. The idea of the living being is also material (and not Platonic) but it cannot be described by physical theories about independent objects since these theories treat objects independent of their context while information can only be understood through a contextual whole-part relationship.

We habitually describe macroscopic objects using concepts, regardless of the matter from which they are built up of. For instance, objects of many different shapes, sizes, and materials are called *tables*. The idea of the table is logically prior to the realization of that idea within matter. In the case of biological beings, the notion of living beings can similarly be logically prior to their realization in matter. By interpreting the wholes in terms of concepts, it is possible to formulate a newer understanding of how the biological beings emerge. In this understanding, it is not necessary to suppose that the parts are created before the wholes. Rather, it is also possible to suppose that the wholes exist as concepts even prior to the parts.

In the previous section we saw how the idea of natural selection is incompatible with physics and mathematics because selection requires the use of collections in addition to parts, which in turn requires the use of second-order logic, and current physics only uses first-order logic. Adding natural selection in current science therefore implies going beyond current physical theories in ways that are quite fundamental. Specifically, we cannot create a second-order theory of nature unless we revise our notions about matter, and we cannot revise our notions of matter unless we solve the problems of incompleteness in mathematics. Additionally we now saw that the idea of random mutation is inconsistent with quantum physics because random mutations assume the existence of independent parts that form wholes and this idea cannot be reconciled with atomic theory. There isn't, therefore, a set of *a priori* real parts that we simply discover. Rather, we must recognize that there is a whole that is divided into parts using different methods. The assumption that there are *a priori* real parts is itself false. The random combination of these parts is also therefore false. Wholes are not created by randomly aggregating parts. Rather, parts are created by dividing the wholes. This division represents an eigenfunction basis in quantum theory and it can be picked up by measurement. The division of the whole into parts is therefore not a random event.

The idea that the random aggregation of parts produces functional wholes creates inconsistent or incomplete theories in physics, mathematics, and computing. This view of biology and evolution is based on classical physical ideas of nature and predates the developments in the 20th century. If these developments are true then biological ideas about random mutations and natural selection are false. The theory of molecular evolution is a heuristic panoply of incoherent ideas such as classical physical notions about causality without the associated determinism, feedback loops within linear theories, a language of functions to select but a language of things to mutate. A closer look at the structure of the theory and the types of arguments it uses reveals it as incoherent and inconsistent.

Game Theory and Altruism

Morality was foreseen as a problem early in evolutionary theory. Evolutionists saw that many animal species act altruistically even to their own detriment. If the "intent" of the gene is selfish, then how does the altruistic behavior come about? The behaviors of collections of rational individuals are often modeled using Game Theory (GT) and this has been brought to bear on the evolutionary genesis of altruism as well. A fundamental premise in GT is that every individual acts for maximizing their payoff, and so evolutionary game theorists want to determine if the selfish payoff may sometimes be maximized by acting altruistically. In general, if each player knows the opponents' alternatives, GT shows that every game settles into one of two main possibilities. Either the population in a group is stabilized at a Nash Equilibrium[14] where no player is benefitted by unilaterally changing their strategy. Or, if the game has not settled into equilibrium, then the evolutionary process actually favors the selfish. Therefore, if actors are entirely rational then there is no incentive to be altruistic in games with full information available.

Of course, GT assumes that the actors are rational, that they can know the strategies and payoffs of the other players, and that they can have intentions to act selfishly or altruistically. These in turn require the existence of meaning, intentions, and beliefs, which in turn pose serious problems for current mathematical and physical theories. We have encountered these problems earlier, so I will not dwell upon them again. While GT rests on certain basic assumptions about the living beings which cannot be taken for granted, even if these assumptions were granted, natural selection and random mutations cannot give rise to altruistic behaviors according to GT, because GT shows that altruism can never be a viable strategy if the group of individuals as a whole is selfish. The only time altruism works is if all or most of the players in the game are altruists. However, that begs the question of why the players are altruistic to begin with? After all, the premise in evolution is that altruism improves the selfish payout, which means that the players must all be selfish to begin with. In such a selfish group, how can altruism emerge unless the player population becomes altruistic at once?

While altruism definitely increases the payoffs for individuals, altruism can dominate in a group only if the majority is altruistic to begin with. The payoffs for the majority are definitely increased by being altruistic and the minority also sees that being altruistic increases their payoffs. When the majority is altruistic, they will reward the other altruists and punish the selfish. This follows from the recognition (confirmed through repeated simulated measurements of multi-player game strategies) that the Tit-for-Tat (TFT) strategy is indeed the best strategy for long-term game payoffs.

TFT reinforces the everyday intuitive notion that if some actors are altruistic then others will also be altruistic towards them. If, however, they are selfish then others will also be selfish towards them. Thus, the negative consequence of a selfish act ensures that over time altruistic actions are preferred. If the majority is altruistic then the entire group tends towards altruism. But, for this to work, the majority in a group must be altruistic to begin with, for which there is really no explanation. Assume for the moment that each member in a group is punishing the other members for their past selfish acts. How will any member ever act altruistically in this scenario? To act altruistically, the player must see an example which shows that being altruistic gets you better payoffs. But if everyone is acting selfishly, then no one is going to set a good example for the others. Indeed, any altruistic actions in a selfish group will look silly because altruists will be exploited by other selfish members. An altruist can only be rewarded by other altruists. So, for any altruism to exist at all, multiple such altruists must materialize at once.

Evolutionists claim that natural selection can act on groups that are altruistic versus groups that are selfish. But how can an altruistic group emerge when the payoffs to an altruist within a group of selfish players are always negative? If nature begins in selfishness, then a single random mutation of altruism will be quickly eliminated. A number of altruistic players must emerge at once, which is very unlikely if the entire group is selfish. For such a group to survive, they must be altruistic to each other but selfish to others. A random mutation that creates a large group that can survive based on altruistic support from other altruists is highly unlikely.

To explain altruism, we have to assume that in any altruistic society most players will be altruistic and only a few will be selfish. The altruistic can then evolve TFT mechanisms to keep the selfish at bay. But as the selfish increase, altruism rapidly declines. Everyone now acts selfishly and everyone is worse off as a consequence. GT has the potential to explain how a group—in the case of a majority—tends as a whole either towards altruism or selfishness, improving everyone's life or worsening everyone's life. But GT cannot explain how these trends are reversed—e.g., why a selfish group will become altruistic or why an altruistic group will become selfish, if everyone in that group has the same basic choices, strategies, and payoffs. That means, if we begin with selfishness, the theory cannot explain how a species ever tends towards altruism over time.

GT is a theory of intentions, but it does not take into account individual *personalities* which will act altruistically or selfishly even if the payoffs are bad. Personalities don't change quickly—i.e. over a couple of bad payoffs. People with altruistic tendencies think that they may have had a few bad payoffs, but there is still merit in continuing on that path. The selfish too rely on the fact that they would not be caught in the selfish act and that they can cloak it long enough not to be caught. GT is inadequate in the sense that it takes individuals to be rational agents who will compute their payoffs at every move. But individuals are not rational in that sense. They have more or less fixed strategies. A lion will not stop hunting even if the rest of the animal population decides to periodically sacrifice one from their group. Note how both the lion and the rest of the animals are better off in this case:

the lion doesn't have to worry about hunting and the other animals don't have to live in fear. But this doesn't happen, because every animal has a certain nature. The lion will hunt; it is part of what it *means* to be a lion—an aggressive killer.

GT builds a theory of evolutionary behavior based on choices, but discards the very framework in which these choices take place—namely, mind and meaning. As we saw earlier, rationality has different meanings in mathematics and in the everyday world. Mathematical rationality is incomplete because it cannot deal with names and concepts—which in everyday language are necessary to deal with the world. Indeed, an animal cannot know what the other animals are doing without using names and concepts. But names and concepts cannot be added to the physical world without recognizing that there is a mind. This recognition is based not upon practical experience with minds, but purely on considerations of rationality; rationality without names and concepts is incomplete. Therefore, there cannot be a complete rational theory of choice without a role for the mind in it. However, if we acknowledge the role for the mind, then we must also recognize that this mind is not always rational. We have personalities and preferences that go against reason. Animals have tendencies due to which they act impulsively.

GT models a player as a rational being without a personality; the player is not *a priori* predisposed to act in a certain ways. The truth is that we all have predispositions to behaviors which determine whether we will be altruistic or selfish. These predispositions make a player belong to a specific species. Different personalities can be connected in an ecosystem that is designed to accommodate such variety. The player doesn't have to act more altruistically or more selfishly than what its nature dictates. The ecosystem as a whole works not because a species is particularly altruistic or selfish, but because it has the right intentional relations of give-and-take with other species. Evolution and GT therefore lead us to similar conclusions. Evolution cannot explain altruism unless cooperative effects happen in different species simultaneously. Game Theory also doesn't explain altruism unless we start out with a large altruistic population. Both theories indicate that individual mutations cannot explain the emergence of large-scale behavior, simply because they cannot overcome the resistance of the rest of the ecosystem. To solve these problems, there is need for an ecosystem model that describes the evolution of the ecosystem as a whole.

This cause is counterintuitive in current science, because current science still lives in the legacy of classical physics where the causality lies in the parts and not in the whole. Indeed, as we saw earlier, classical physics does not even treat the whole as a real entity. A different type of whole realism is indicated by quantum theory in which the macroscopic objects could be seen as real ensembles and the parts in that whole are created by an observer's actions. Such a theory will also imply shifts in notions of causality. In particular, causality is not that which emerges bottoms-up (atoms to macroscopic objects). Rather, causality is the evolution of macroscopic objects, which in turn determines the states of atoms. It is now possible to speak about the evolution of ecosystems without trying to describe the ecosystem's 'emergence' through random mutations.

The Implications of Meta-Biological Considerations

The readers would have noticed that I have sidestepped *biological* arguments to criticize evolution. I have rather used arguments from physics, computing, mathematics, and game theory to critique shortcomings in the current evolutionary model to explain diversification. These arguments are briefly recapitulated again:

- A stable ecosystem subsists in a cycle of interrelations and any change in this cycle ripples throughout the entire ecosystem. Unless the ecosystem changes at once, the changes are not sustainable. Random changes in an ecosystem will be rejected because the ecosystem as a whole is not ready to change.

- The idea that the world is built up of *a priori* real atoms is a classical view of nature inconsistent with quantum theory because how quantum probabilities are overcome to form stable objects seen in observations is still unexplained.

- The quantum explanation of chemical reactions is causally incomplete and cannot account for random mutations. Quantum theory injects a variation of the Zeno's paradox where every action requires a prior action, ad infinitum, which makes the causality incomplete. This explanatory gap requires a new theory where possibilities are converted into reality. However, with such a theory, there is no room for randomness in nature.

- A closer look at statistical mechanics reveals that the ideas of reduction and non-linearity are contradictory. If reduction is true then biological systems must be reversible and linear. That in turn entails that life will not have a time direction and living beings would violate the second law of thermodynamics.

- The problems of indeterminism in relativity illustrate a type of matter redistribution transformation that is *what* deterministic but *who* indeterministic. The solution to this problem needs a role for choice where objects are distributed over trajectories and in which there are many possible distributions in nature, which can only be chosen by observers because the theory of matter is consistent with all possible matter distributions.

- The existence of meanings in the mind requires a new theory of matter when the mind is reduced to the brain. Classical physical views are inconsistent with the existence of meanings in the brain. Of course, they are inconsistent with the existence of meanings in books, pictures, and in scientific theories themselves. However, the problem of meaning in the brain makes this problem much more relevant and immediate to biology.

- Paradoxes such as Gödel's Incompleteness clearly demonstrate that a physical theory cannot explain meanings because the induction of names and concepts into a theory of things or physical tokens leads to logical inconsistencies in mathematics.

- The adaptation of Turing's Halting Problem to biological functions shows that random mutations will produce eternal beings inconsistent with observation. Nature can prevent such eternal beings only if it only produces meaningful programs, which requires meanings to exist prior. The idea that nature produces random programs leads to a paradox such that only random mutation or natural selection could be true but not both.

- The ideas of natural selection and random mutation involve the use of two different types of languages—natural selection needs an object language while random mutation needs a property language. These languages can only be combined by the use of second-order logic, but there cannot be a consistent and complete physical theory that employs second-order logic.

- The existence of altruism can only be explained if nature begins in altruism. Game theory predicts that Tit-for-Tat strategies are the best winning strategies. Therefore, altruism can exist only if the majority of the population is altruistic. In the case of evolution, this implies that nature would have to begin in altruism.

It follows that if biology is built on the idea that living beings are epiphenomena of physical states then the theory of evolution can either be consistent or complete: if either random mutation or natural selection is used then the theory is incomplete; if however both random mutation and natural selection are used then the theory is inconsistent. Furthermore, ideas about probability and indeterminism are inconsistent with the reality of atoms and molecules. Finally, such living beings cannot have meanings and experiences.

These problems are, however, *theoretical* and not those of experimental observation. Experiments show that the brain is a stable object, although quantum theory does not explain its stability. Experiments also show that brain events are correlated with mental reports of meaning and experience, so it is fair to assume that the brain is indeed a representation of the mind. Since the brain works without creating a logical contradiction, reality must be able to store meanings. Observations reveal that living beings have a finite lifespan implying that nature must somehow produce only meaningful programs. The problems are thus not in the reality or its observation. The problems are in the current theoretical *explanation* of that reality and observation. To correctly explain how the macroscopic world arises from atomic theory, and to explain how these objects signify meanings without creating logical contradictions, we require new *theories* of nature. In such theories, the whole will have a reality distinct from the parts that make it up. Both the wholes and the parts would be associated with meanings. The evolution of matter would be modeled as an evolution of meanings, and its causality will be based upon a different notion of cause and lawfulness. With such laws, it would be possible to speak about how ecosystems as a whole can evolve according to laws without random mutations. Such a theory, however, needs to overcome not just the randomness of current evolutionary theory but also the indeterminism in other fundamental theories in science. This is a far larger shift in science as a whole, than merely an adjustment of evolution within biology.

What Real Evolution Looks Like

The question of the diversification of the species is now not a question of evolution in the classic sense of mutate-and-survive. It is rather a question of how many different types of living beings exist in a system and must co-evolve to make any small change in the system benefit every member in the system. Without a concerted evolution of various stakeholders in an ecosystem, unilateral changes will only be detrimental to the evolving species, because other species that do not benefit from such changes will evolve in ways opposite to the unilateral change, to nullify the advantages to the benefitting species. To know how a system can co-evolve collectively, we must know how many types of species exist within a system. This in turn implies knowing all the DNA types in an ecosystem to make the ecosystem as a whole a viable system for all the players. An ecosystem circulates meanings (e.g., useful things) like an economy circulates products and services. The survival of a species depends on how well it fits vis-à-vis other members of the ecosystem. In a stable ecosystem, no species can change unilaterally and expect to have a better chance of survival because changes to the consumption patterns of a species will affect other species and these effects can ripple back to the original species, often as a retaliation to change.

The only way a large ecosystem can evolve is if all the stakeholders (or at least the majority of them) are collectively benefitted by the transition. But, now, it is no longer an individual's mutation. Rather, the ecosystem should evolve as a whole. This puts the evolutionary and ecological approaches to diversification at odds. In evolution, many species are created from a single species because they mutate and diversify, and are selected by the environment. In an ecosystem, all species *co-evolve* as no individual has a better chance of surviving unless other species are also going to cooperate.

The problem of many species is not about how they emerged from a single species, but pertains to knowing how an ecosystem can be *logically* built out of different species. An ecosystem must comprise only as many species as the ecosystem needs. Adding or removing a species has drastic consequences that cannot be reconciled unless the ecosystem as a whole evolves to a new state.

To understand ecosystems and their collective evolution, a new kind of science is needed in which a system as a whole evolves in its entirety rather than in its independent parts. This, in turn, requires the recognition that there is indeed some whole aside from the parts, an idea that I argued earlier is supported by the problems in physics, mathematics, computing and game theory. In the collective evolution, objects do not evolve by their own physical properties. Rather, the ecosystem as a whole evolves and forces the individual to adapt to it. The present evolutionary model extends the causal model of individual change into a change to the ecosystem, and the new evolutionary model will extend the evolution of the whole into the evolution of its parts. With such a theory of evolution, is easy to see how nature is governed by laws that deal in more and more abstract conceptual objects rather than the smallest physical particles in nature. When nature is defined as concepts, then its evolution is governed by laws of meaning rather than forces of nature.

This idea has a remarkable resonance with the evolutionary data. Fossil data clearly shows many gaps if evolution were to take place through incremental changes. What is consistent with the data is the idea that biological ecosystems hop as a whole from one stable state to another. We cannot explain that kind of change using a perturbation theory of random chance mutations. Unilateral random perturbations—even if they occur—have to be reversed because otherwise the random mutant will be eliminated. At best, therefore, the random mutant population will oscillate and return back to its original state or would be eliminated by that change. Random mutations will thus kill the mutant species if the inverse mutation does not quickly reverse the changes. The only way a change can survive is if all the species collectively change. But that cannot be called a random chance mutation. Fossil data suggests the existence of collectively consistent ecosystems which hop from one state to another. That data is inconsistent with random mutation.

The next chapter discusses a new material foundation that solves the problems of incompleteness, indeterminism, uncertainty, and logical contradiction in mathematics, computing, and physics. I believe that it is very important to understand the problems of science in a broader context before we see their significance in the context of biology, especially since biology is and will be based on more fundamental theories of nature. While biology will deal with macroscopic objects and living ecosystems, its theories and assumptions cannot contradict fundamental logical, mathematical, and physical principles. Only when we have understood the need for a new theoretical foundation in computing, mathematics, and physics, can we understand their role in the context of biological evolution. However, once we have grasped the fundamental tenets of the new foundation in science, it will be easy to see their implications for biology. If meaning and mind have a fundamental role in nature, then living beings are representations of ideas originating in the mind. The evolution of the body is prior an evolution of the mind.

3

Matter and Mind

In the world of the very small, where particle and wave aspects of reality are equally significant, things do not behave in any way that we can understand from our experience of the everyday world...all pictures are false, and there is no physical analogy we can make to understand what goes on inside atoms. Atoms behave like atoms, nothing else.

—*John Gribbin*

Phonosemantics and Quanta

A deeper understanding of how matter encodes meaning can begin with the understanding of how meanings arise in natural language expression. This is simply because language is also material, especially if we think of language as sounds and texts that embody meanings. If we look closely at language, we find that meaning gradually grows from alphabets (which don't seem to have meanings) to words, sentences, passages, and entire texts (which seem to supply ever increasing amounts of meaning). For instance, while English alphabets themselves don't seem to have any meanings, when these alphabets are aggregated into phonemes they already begin to denote some meaning. Studies in phonosemantics[15] have demonstrated that words beginning with similar phonemes denote similar *classes* of meaning. For instance, the phoneme /str/ partakes in words like 'string' and 'strap' which both denote something long and thin which can be used to tie things or hold them together.

When phonemes are aggregated into words, the meanings become more complex. There may be some ambiguity about whether a word denotes a noun or a verb, and that is generally resolved when words are combined into phrases. The meanings of complete phrases may be ambiguous, and these ambiguities are resolved when phrases are combined into sentences. And this pattern continues almost indefinitely. For instance, the meaning of a sentence is often fixed by the sentences that precede or follow it. Often the meaning of an entire passage is understood by taking into account the meanings of the passages before and after it. In thriller movies and mystery novels, the full meaning of a statement or event made early in the story becomes clear only at the end, when a number of additional facts have been revealed and their order is understood. Although this gradual development of meanings from meaningless letters is obvious in language, we still don't understand how it comes about. This has deep analogues with the fact that we don't know how a collection of physical particles can encode meanings.

One way to solve this problem is to redefine language not as the written word but as the spoken words (i.e. phonemes). Phonosemantics asserts this idea as well. For instance, the similarity between the meanings of words is not based on how they are written, but how they are *spoken*. Thus, the meaning of the sound /kæ/ as in 'cat' is different from the meaning of the sound /tʃæ/ as in 'chair'. Both 'cat' and 'chair' begin with the same letter 'c' but they belong to different semantic categories. This could be understood if we built language not from alphabets but from phonemes. In the above example, 'cat' and 'chair' use different phonemes. Through the idea that meanings arise from how language is spoken rather than how it is written, some dramatic shifts in the understanding of language can take place. Now the fundamental units of language are not meaningless letters which magically produce meanings. Rather, the units of language are phonemes which seem to have meaning.

Of course, why phonemes have meanings is still an unresolved problem. To understand how this problem can be handled, let's look at how tones of sound become notes in music. A tone of sound is, physically, just amplitude, frequency, and phase. Indeed, the tone of sound is not a note by itself unless we comprehend multiple such tones collectively. When tones are heard collectively, they constitute musical *scales*. Within the scale, each tone also forms a *note*. The succession of such notes forms a composition which denotes various kinds of meanings and emotions. In Indian classical music, for instance, different scales and note orders are used to depict times of day such as morning, afternoon, and night, seasons such as spring or winter, social occasions such as birth, marriage, and death, moods and emotions such as happiness and sadness, etc. Individually, each tone is meaningless. But, collectively, the tones are created to express a meaning. Outside the scale, the note is just a tone. But, within the scale, the same tone also symbolizes a note. The meaning associated with the tone is tied to the tone but it cannot be understood unless we account for the interrelations between various tones.

This idea can be extended to the meanings of phonemes in language as well. Each phoneme is a sound pattern and can be described physically as some amplitude, frequency, and phase. However, the meaning of the phoneme will only be understood in relation to the other phonemes. This relation may differ in the native speakers of different languages, or even sometimes in the abilities of speakers of the same language, although the majority of speakers intuitively understand this compass of sound-meaning relations.

A prominent example of this fact is demonstrated by experiments on the Bouba-Kiki effect[16]. The sounds 'bouba' and 'kiki' have no dictionary meanings associated with them, which means that if we thought of language primarily as a written medium of communication, then the sounds 'bouba' and 'kiki' would be meaningless. However, if subjects in an experiment are shown two shapes—one curvy and bulbous while the other sharp and jagged—and asked to associate these shapes with the sounds 'bouba' and 'kiki' then about 95% of all subjects associate the sound 'bouba' with the curvy and bulbous shape and the sound 'kiki' with the sharp and jagged shape. This observation illustrates that we have an intuitive understanding of the *relation* between the words 'bouba' and 'kiki'—as we associate them with curvy and jagged shapes—although the words 'bouba' and 'kiki' themselves have no predefined dictionary meanings.

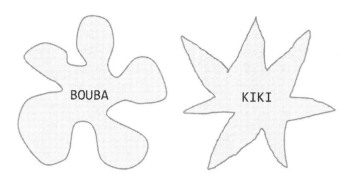

Figure-15 The Bouba-Kiki Effect

The problem of meanings cannot be solved by associating each word independently with a meaning—which is what most current attempts at understanding meaning endeavor for. The problem of meaning is rather the relation between objects within a collection, since it is this relation that makes the objects symbols of meanings. If we think of the objects as physically independent things, we can never demystify how these things become symbols of meanings in a collection (a collection after all is just a random aggregation of independent parts). Also, any notion of individual objects that tries to define these objects simply in relation to each other never works because of the problem of recursion: object X must be defined in relation to object Y, which must be defined in relation to object X, and so forth. When the definitions of both X and Y depends on the prior definitions of Y and X, then neither object can be completely defined. Recursion entails that the system is again logically incomplete. To address this problem, a conceptual shift where the parts are created by dividing the whole is necessary. Each part is still a physical thing, but the whole is logically prior the parts. If the parts are produced by dividing the whole, then both the whole and the part can be symbols of meanings. This in turn changes our outlook on physical things, and the space and time in which they exist.

All physical entities are objects. But collectively they become symbols. When they are aggregated, there are two ways to look at these entities. First, we can measure their physical properties like the frequency, amplitude, and phase of a sound wave. Second, we can see them as symbols of meaning. The meaning in the sound is produced due to the *interrelation* between the whole and the part. This viewpoint changes the landscape of how the questions of meaning are approached both in language and in scientific attempts to encode meaning in matter. The fundamental insight is that meanings arise *automatically* when collections are divided into parts.

Within a collection, the individual objects become *types*. A complex space-time arrangement of such types can denote complex meanings. In classical physics, everything is of one type—particles or waves—and their aggregations still remain the same types.

To describe meanings in matter, a theory of nature must describe the parts of a collection as different *types*. Classical physics cannot be that theory because the classical description is complete (although inconsistent with observed facts). Quantum theory can be adapted to become a theory of types, although the present theory still describes things and that description is incomplete. Note that a different type of thing is also a different thing, although a different thing is not necessarily a different type of thing. In that sense, the type distinction is stronger than the thing description, and the type description can be used to complete quantum theory without changing the observables. The shift would be similar to how a tone can be described as a note, or a sound as a representation of a meaning.

The type description of nature alters some fundamental ideas in classical physics. One such idea is that of motion. As I mentioned in the introductory chapters, in classical physics, when a particle moves it has a different position, but it is supposed to be the *same* particle. In quantum theory, this is no longer true, because particles at two different positions represent two different eigenfunction states, and each such state denotes a different particle. Therefore, when a quantum particle 'moves' from one location to another, we cannot claim that the *same* particle changed position as was the case in classical physics. We must rather say that one particle transformed into another particle. The particles at the initial and final positions are different particles, not just different positions of the same particle. Indeed, the particle is not moving in the classical sense; it is rather being transformed into a different particle.

Current quantum theory carries forward the classical ideas of motion. In the current theory, differences in position are differences in *states* rather than differences in *identity*. Therefore, when a particle jumps from one state to another we do not say that one particle disappeared and another one appeared. We only say that the *same* particle jumped to another state. This vestige of classical thinking leads to problems when the jump has to be explained, because the theory cannot explain when, how, and why the particle jumps.

If these particles were treated as symbols of meanings, then each location of the particle would denote a meaning. Since distinct positions help distinguish the particles, and if the position denotes a meaning, then the particle at a different position is also a different *type* of particle. The eigenfunction state of the particle (which is currently described as amplitude, frequency, and phase) can now also be described as the meaning associated with a phoneme. The succession of particle states (which classically looked like the motion of a particle) will now represent a succession of phonemes, analogous to a sentence. While the succession of particles in current quantum theory cannot be explained, this succession can be predicted when the quanta are treated as symbols of meanings. The succession will now express the same meanings that were earlier encoded during state preparation. State preparation and observation are now procedures to encode and decode meanings. Unlike classical physics, these procedures now represent parts of a communicative act. Essentially, state preparation encodes some meanings (perhaps over time) which the observation decodes (again, perhaps over time).

When a collection of quantum particles is treated physically, the problem of the statistical collapse of the wavefunction follows whereby a state randomly emerges from probability. The collapse hypothesis requires something to collapse the wavefunction, which is outside quantum theory. Attempts to solve the quantum problem today are still focused on understanding the atomic world in terms of the classical pictures of nature. A collection of quantum particles is still treated as a collection of different things, not of different types. Could it be that the quantum reality involves a role for viewing things as types? If this were the case, then we could say that particles at two different positions are different *types* of particles. The state of the particle could denote a type, which is interpreted as a representation of a meaning. Such a typed theory of reality can then naturally incorporate ideas about the mind without requiring an extraneous role for an agent that collapses the wavefunction.

A Semantic View of Reality

There have been prior attempts to incorporate a role for the mind in quantum theory. John von Neumann, for instance, postulated a role for consciousness in collapsing the wavefunction, but this hypothesis did not explain the *mechanism* by which such a collapse would occur. Without such a mechanism, quantum theory could not explain how a stable reality emerges from a cesspool of probability. The approach also maintains a mind-body dualism where the mind collapses the probability into a reality. This dualism is not useful because the mechanism of mind-body interaction is not known. Empirical observations suggest a theory in which the mind is represented in matter itself, not as an external agent to collapse probabilities. In what follows, therefore, I will discuss an approach to quantum theory that explicitly connects questions of quantum reality to the idea that meanings are encoded in matter. What follows is a summary of what I call the *Semantic Interpretation of Quantum Theory* (SIQT). I have described this interpretation elsewhere[17] illustrating how it not only solves the conceptual difficulties in quantum theory but also helps us see how a new theory could overcome the probability and indeterminism. The summary here will only capture the points relevant to subsequent discussions about evolution.

Let's begin by understanding the basic conflict between classical and quantum theories vis-à-vis how they individuate objects. In classical physics, each stationary particle has a unique location in space and particles are physically distinct entities because of that location; if a stationary particle has a unique location, then it must be a unique particle. Of course, the particles may not be stationary. In that scenario, different particles may occupy the same location at different points in time. Therefore, location serves to distinguish stationary particles but not particles in motion. However, even moving particles can be individuated by a combination of space and time coordinates. This fundamental idea from classical physics is violated in quantum theory where multiple particles can have the same location even in a stationary state (i.e. when the particle state does not change in time). But if multiple particles have the same position in space then how can we claim that they are indeed different particles? The reason is that when we measure these particles, they arrive at the same location at different points in *time*. The difference in time makes us believe that they are indeed different particles. Now, this might seem like the classical case where different particles can have the same location at different times. The difference, however, is that quantum theory does not *predict* which location will be occupied at which time, something that classical physics could do.

In classical physical theory we can distinguish particles based on space and time. In quantum experiments, too, we can individuate the particles based on space and time. However—and this is the key point—we cannot individuate the particles in space and time using quantum *theory*. The theory is incomplete because it does not predict everything that we can observe. We observe that different particles arrive at the same location at different times. Quantum theory only predicts the location but not the time. That is, we cannot predict which particle will arrive where *and* when. We can predict (in one viewpoint[18]) *which* particle will arrive *where*, but not *when*.

If both space and time are needed to individuate a particle, and the theory only predicts the space but not the time, then the resulting prediction is probabilistic. Most quantum interpretations treat this probability as a feature of reality arguing that nature is not deterministic. But this view creates conflicts between the atomic and the macroscopic worlds. If reality is indeed probabilistic—as quantum theory tells us—then objects should randomly appear and disappear. Why is the macroscopic world then *stable*? Why doesn't it randomly appear and disappear such that when I wake up every morning I only have a certain probability of seeing my hands and legs in the same place where they were last night? This paradox is called the quantum *Measurement Problem* and it remains unsolved. Pending the solution to this paradox, quantum theorists selectively apply quantum theory to the atomic world and not to the macroscopic world. Of course, quantum theory is a theory of reality, not a theory of atomic reality. But unless we can solve the measurement problem, its application to the macroscopic world is riddled with problems. The idea that the macro-scopic world is classical and deterministic therefore lies at the root of the quantum problems.

SIQT argues that the quantum is an atomic vibration that can be viewed as a *phoneme*. The quanta, quite like the phonemes, have physical properties. But measuring these physical properties is not the correct description of the phoneme. The correct description must treat the phoneme as a *symbol* whose meaning is given in the distinctions between the quanta. Therefore, the quanta are not defined independently as classical particles were. Rather, a quantum is defined in relation to ensemble of quantum particles. Accordingly, the state of the quantum is intrinsically linked to the state of the ensemble. We can no longer treat these particles like classical particles. Rather, we must treat them as symbols of atomic meaning.

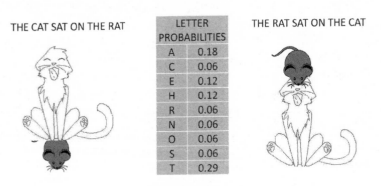

THE CAT SAT ON THE RAT

LETTER PROBABILITIES	
A	0.18
C	0.06
E	0.12
H	0.12
R	0.06
N	0.06
O	0.06
S	0.06
T	0.29

THE RAT SAT ON THE CAT

Figure-16 Quantum Probabilities Hide Meanings

This viewpoint facilitates a better understanding of quantum incompleteness. A quantum ensemble is now like a book. The position of the quantum denotes the *type* of the phoneme and the arrival of quanta in time denotes a specific *instance* of that type. The succession of quantum arrivals therefore represents a succession of symbols as in a book. If we reduce the book to the physical properties of its individual symbol-tokens, we will measure token probabilities instead of meanings. If you read a book without knowing the language, you can measure letter probabilities and conclude that letters recur with some well-defined probability. Your measurement cannot lead you to the conclusion that the book has meaning and thus the most important fact about the book (perhaps the very reason why it exists as a book) will escape you. Current quantum theory measures phonemes like the counting of letters in English. The letters have no separate meaning, and if we measure letter probabilities we cannot predict the succession of letters in time. The succession of symbols represents a new kind of order when quanta are seen symbolically. This order represents the meaning in the book, and the succession of symbols can be combined to produce a complex meaning if symbols themselves have meanings. The succession of symbols is meaningless only when symbols are meaningless.

The symbolic properties are however not limited to atomic reality. Everyday objects like books, music, and paintings are also physical things but they cannot be reduced to classical particles. All macroscopic objects can now be seen as instances of some concept. A symbol combines the physicality of an individual and the meaning of a concept. Current quantum theory describes only the physicality which is given independent of all other particles in the ensemble. To study semantics, science must treat the particles collectively as part of some ensemble, such that the collection gives meanings to parts.

The remarkable feature of this view is that it is already consistent with our macroscopic world (although not consistent with the classical physical description of the macroscopic world). The idea of symbols and information is derived from our prior experience with macroscopic reality. Such a view also seems to explain the observed statistical nature of quantum theory as not a feature of reality itself but that of a *theory* that treats symbols as things. Specifically, by looking at symbol frequencies we will not understand meanings. The symbols that seem to appear randomly one after another are not actually appearing randomly. If only we knew the language in which nature encodes information, we would also be able to decode the symbol sequences into propositions of meaning.

The semantic view therefore offers new predictions and phenomena that go beyond current quantum theory. Current quantum theory is like measuring the weight and length of a book while a new quantum theory will actually read the book. Physical measurements on the book are not false. But they constitute an incomplete description of nature. The difference between quantum and classical theories is therefore not that one is deterministic and the other probabilistic. The difference is that classical physics measures properties of objects in isolation while quantum theory describes properties of objects in collections. Classical physics described each particle as being independent of the other particles. Quantum theory shows that a quantum's properties are *defined* in relation to other quanta in the ensemble. This disruption in thinking requires a more dramatic shift from classical physics than current quantum theory makes. The shift is that classical physics was a theory of things while quantum theory is a theory of symbols. A symbol is also a thing, but a thing is not necessarily a symbol. In that sense, quantum theory is a more generalized description than classical physics.

The so-called transition between classical and quantum theories can therefore be constructed in a new way using semantics if we treat quantum objects as symbols and classical particles as objects. The transition would be that if we study collections, we can decode meanings but if we study isolated objects we can only perceive physical properties. The surprising aspect of this interpretation is that the theory of collections is a more fundamental theory than the theory of the objects. But if we recognize that there are no classically independent particles in the universe, and everything exists in ensembles this is not at all surprising. In the simplest case of observation, there are at least two objects—the observer and the observed—which can only be defined mutually and not independently of each other. What classical physics describes as independent particles is therefore an idealization. This idealization was false even for macroscopic semantic objects, but science ignored it under the belief that matter exists independent of the observer, and since the observer is another material object, material objects must also exist independent of the measuring instruments. The problems of this view can be solved in a theory that describes ordinary objects themselves as tokens of meaning arising in collections. With this approach, we can resolve the problem of symbol probabilities.

Of course, quantum problems do not end with probabilities. A further, deeper problem is that quantum theory is consistent with many possible matter distributions in space-time. A distribution is a way to partition some energy into objects. Quantum theory indicates infinite ways to divide a fixed total amount of energy and matter, thereby creating different sets of objects. These distributions are called eigenfunction bases in the theory, and they stipulate different positions and momenta for quantum particles. We have already seen how problems of quantum probabilities are solved by treating quantum objects as symbols. From here, it follows that each matter-energy distribution should be viewed semantically as a different encoding of meaning in the same total matter-energy (or a different physical representation of the same meaning). Quantum theory implies that once we collect an ensemble of particles, we must deal with the whole ensemble, and there are many ways to distribute matter in that ensemble, keeping the total energy constant. The theory, however, does not predict which distribution is real. This indeterminism in quantum theory is even deeper than the probabilities associated with the predictions of various particles within a certain basis because the theory does not even probabilistically predict which of the matter distributions is more likely.

In the semantic view, this quantum indeterminism is the ability to encode different kinds of information in matter, keeping the total energy and matter remaining constant. Intuitively we know this as the ability to write different books using the same total ink and paper. Or the ability to encode books, movies, music, or art in the same computer hard-drive without changing the total weight of that hard-drive. If an ensemble has already been prepared, quantum indeterminism implies that there are many ways to express the same meaning, each time using a different set of words and word order. The inability to predict the eigenfunction basis in the theory illustrates that quantum theory is agnostic about the *structure* of matter within an ensemble if we are dealing only with an ensemble as a whole. The theory deals with the total amount of energy and matter in the ensemble, not with the distribution itself. To determine the distribution, we must know the information that was encoded in matter prior to observation. The encoded information can be decoded by observing the physical states of particles, but to encode this information the information itself must exist in some form prior to encoding. The meaning that exists prior to encoding cannot be observed by the senses (taste, touch, smell, sound, and sight), although we must suppose that it exists before being encoded.

The causality underlying that encoding of meaning is thus not observable, but the cause must exist in a non-observable state prior to encoding. Contrast this to current physical notions of causality where both the cause and effect are physically observable entities. The problems of quantum theory imply that the cause of change is not in space-time, although its effects are visible inside space-time.

Quantum Theory and the Mind

Of course, this idea is not new and has existed for centuries in classical theories of mind-body dualism. Dualism suggests that the mind produces and consumes meaning through matter. Ideas that originally existed in the mind can be encoded in matter by some mind, from where other minds can decode them, provided they employ the same language to encode and decode. The main reason these theories could not enter mainstream science is because scientific theories required that the cause be empirically observable.

As philosopher of science David Hume said, to establish a causal connection between objects we must be able to observe both the cause and effect and their succession consistently and repeatedly. Hume's view of causality is good if causality has to only explain the effects of one object on another. This view, however, does not explain how objects themselves came into existence. If objects have to exist to cause other objects to move, then there must be some *a priori* distinct objects, which require the universe of particles to be always *a priori* real. Quantum theory refutes this idea because in quantum theory particles can emerge from a vacuum and disappear into that vacuum. This appearance and disappearance has no causal explanation, and quantum theory stipulates that this fuzz of space-time-energy is randomly creating and dissolving particles. Randomness from the physical theory percolates into biology as the notion of random chance mutations, where molecules can mutate randomly to create new molecules. But, as we saw earlier, the idea that nature is randomly creating and dissolving objects is inconsistent with the observation of stable objects in the everyday world. The randomness in the theory must be viewed as our current inability to predict the order of events, because the events signify meaning.

In the semantic view, meanings originate in the mind and are embedded in matter, creating objects as symbols of meaning. If we know the meaning we can predict the order in events. But if we don't know the meaning, and we see these events as physical properties, then there is no theoretical explanation for randomness.

The nature of randomness in quantum theory reinvigorates the role of the mind within quantum theory. In this role, the mind does not collapse the wavefunction possibility into reality as von Neumann's interpretation claims. The mind rather creates a reality which current quantum theory describes as possibility. Possibilities are a limitation of the current physical description of nature, and this limitation can be overcome when reality is described as symbols instead of things. The mind creates a matter distribution by adding contingent ideas to abstract ideas. The material objects are contingent ideas while the mind is an abstract idea: the meanings of the symbols are given in relation to the mind, and these meanings can only be perceived by the mind. Through this encoding, a mind communicates with other minds, because once meanings are encoded as objects they become accessible to other minds. Matter distribution comprises the objects, but also their locations, durations, and directions. The meanings in the mind are encoded as 'vibrations' in space-time. We might paraphrase this view by saying that the mind externalizes its ideas as vibrations which become phonemes.

A remarkable fact about quantum theory is that it describes all things in terms of packetized waves. We are also aware that these waves can be modulated to carry information—e.g., voice, video, and data to a wireless communication device. Physics, however, describes these waves as phase, amplitude, and frequency although the users of the communication technology interpret the waves semantically as voice, video, and data. The same wave is therefore treated as a *symbol* in communication theory. These symbols are currently meaningless, because physical instruments only measure the properties of individual quantum objects, not the relations between the objects and the complete ensemble—the speaker in this case. The interpretation of meaning is therefore pushed into the mind, although it exists in the vibration of the waves themselves.

The semantic view of quantum theory treats the quanta as representing *semantic* information, which is presently reduced to physical states. The arrival of quantum particles at the wireless receiver therefore does not represent randomness, if the particles are in fact symbols. In the same way, the appearance and disappearance of particles is not random if this dynamic represents meaning evolution. This viewpoint helps us resolve the problems of statistics and indeterminism within the theory and ideological conflicts with the determinism of the macroscopic world at the level of fact and observation. Now physical objects are vibrations like phonemes. The object itself is the word whose physical states denote meanings.

A semantic view of atomic reality shifts the notion of causality from something that exists *within* space-time to something that causes space-time itself. If causality is within space-time it cannot explain how particles randomly appear and disappear in space-time. By supposing that meanings are used to create space-time locations and these locations are quantum waves, the viewpoint can explain the causal anomaly that arises in current quantum theory.

This new idea requires the insight that the classical world is not deterministic in the sense that classical physics conceives it. If that were so, the notion of choices—and what we call consciousness as an inalienable aspect of all living beings—would not be possible. Classical physics is also inconsistent with the existence of meanings that refer to other objects. Fundamentally, every object in classical physics must describe itself and not another object because an object's properties cannot refer to another reality. Classical physics therefore does not explain the existence of the mind. However, even more fundamentally, it does not allow for the existence of any symbolic object—books, music, pictures, and science—in the universe, such that the meaning of that symbolic object is in that object itself. Indeed, if quantum theory had never been discovered, science would still have failed to account for meanings and choices.

There are two paths to address the problem of meaning. First, we might say that books, pictures, and music are not in themselves meaningful and meanings are interpretations created by the observer.

Second, we might say that books, pictures, and music are themselves meaningful and meaning natively exists in these objects although it requires an educated observer to decode them. In the first approach, the problem of meaning is a problem for biology and in the second approach it is a problem for physics. That is, if we say that books themselves are meaningful then there must be some physical explanation of what makes a book meaningful because the atoms and molecules have no meaning in current physics. If, however, books themselves are meaningless and the brain only acquires these meanings, then meanings are some mysterious—yet to be ascertained—property of the brain. This division between the world and the brain gets muddled if the brain is also a physical object. If the laws governing the brain are the same as the laws governing the book, then it does not really help to push the problem of meaning into the brain because the problem still needs to be solved in terms of atoms and molecules. If we claim that brain states can hold meanings then it should be possible for these states in the book to also hold meanings. If the brain and the book are both material objects then they must also have the same abilities to encode meanings.

Overcoming Incompleteness

The formulation of a semantic theory of matter requires a new mathematics in which ordinary language categories such as naming and meaning are compatible with the existence of things. As discussed earlier, current mathematics forbids the coexistence of things, names, and meanings because mathematics cannot distinguish between *is-a* and *has-a* in the theory of numbers. This distinction exists in our common-sense distinction between objects and their properties. The object *is-a* such-and-such and it *has-a* this-and-that property. For instance, the object is a car and it has a steering wheel. Or that John is a man and he has a brother. Here, John and the car are objects but they are either *composed of* other objects or they bear *relations to* other objects. The meaning of a 'car' is either what it is composed of or what relations it bears to other objects. These two kinds of meanings require a distinction between *is-a* and *has-a* in the theory of symbols. With such a distinction, we can speak about the thing—which is the symbol—and its properties which are its meanings. The thing *is a* symbol and it *has a* meaning. Logically speaking, there is a subtle distinction between an object and its properties and yet we cannot separate the object from its properties. That is precisely the kind of distinction needed between symbols and their meanings. The symbol has a meaning, but it is not itself the meaning (i.e. the meaning does not reduce to the symbol).

This logical distinction between things and their properties can be formalized in a theory if the relation between a symbol and its meanings is given through different types of *measurements* on the symbols. Such measurements measure different properties of physical objects, and they derive different meanings. Some measurements analyze parts of an object and they attribute *compositional* properties to the object. Other measurements look at an object's relations to other objects and they attribute *intentional* properties to that object. The compositional and intentional properties of an object are called its meanings. They are associated with symbols, but they are not identical to the symbols. Different classes of meanings—e.g., names and concepts—are also associated with a word, but they are not identical with the word. The relation between a word and meaning is unique to a specific context. The categorical distinctions between different types of meanings must be associated with the various ways in which we can 'observe' the symbol. This means that when we interpret the symbol as a name we are using a different method of observation than when we interpret it as a concept. Therefore, a scientific theory of symbols can be constructed by recognizing different methods of observation which derive different meanings from the same symbol. All methods of observation can be employed simultaneously, and therefore all the meanings of the symbol are concurrently possible although they are not identical.

Such a material view of meanings resonates with quantum theory when the distinction between a symbol and its meaning is equated with the distinction between eigenfunction and eigenvalue. The eigenfunction (symbol) is reality and the eigenvalue (meaning) is produced by performing measurement operations on the symbol. Such a theory of meaning does not reduce the *experience* of meaning to physical properties, although it indicates a method by which these meanings can be *represented* in matter. Essentially, different operators in quantum theory can correspond to semantic categories such as meaning, naming, algorithm, problem, etc. The eigenvalues of these measurements will denote conceptual types that denote the semantic differences within a given type of semantic category.

Quantum operators include the measurement of space-time location and direction. They also include the measurement of phase and frequency of the quantum wave. All these properties are in space-time. Indeed, they are properties of a quantized space-time. By reducing the categorical and type distinctions to space-time properties, it is possible to encode meanings in space-time itself. Of course, now, space-time is not a physical container of things. It is rather a semantic container of types, in which different locations and directions in space-time denote the different types. In a physical space-time, the distinction between objects is a distinction between *individuals*. In a semantic space-time, the distinction between objects is a distinction between *types*. An object at a different location is not just a different object, but also a different *type* of object. The position of the object itself denotes a conceptual difference and if the object has a different position then it must also have a different conceptual type. A different type of object must also be a different object, although a different object is not necessarily a different type of object. This means that the type distinction is stronger than physical distinction, and objects that are distinguished as types will be described as different types although measured as different entities.

The quantum-compatible solution to Gödel's incompleteness implies that distinctions between individual objects cannot create types (as Gödel's Theorem proves) but distinctions between types can create individuals (as the solution to the problem indicates).

The solution to Gödel's incompleteness requires that the symbol and its meaning are not identical (because then the categorical distinction between things and meanings will not exist) and the symbol and its meanings are not separate (because then we will not be able to speak about the meaning of the symbol in the same theory we speak of symbols). These two requirements appear contradictory but they can be reconciled through a distinction between reality and its observation. If reality is treated as symbols, and its various meanings as observations of the symbols, then it becomes possible to distinguish between the symbol and its meaning (due to the distinction between reality and observation) while meanings can be applied to the symbol (as we apply observations back to reality).

This idea is further quantum-compatible if we treat the quantum as a symbol and its measurable properties (position, time, momentum, energy, angular momentum, spatial direction, spin, and temporal direction) as representations of meaning. These representations don't constitute the *experience* of meaning but they can denote the *linguistic* objectification of meaning in matter. To make this idea useful to science, all physical properties being described thus far as quantities must now be described as types. The position of a particle, for instance, must denote a concept rather than a quantity.

The symbolic expression of meaning in matter requires us to treat space-time as a semantic domain. Objects in this domain are symbols whose meanings are given by the object's measurements. However, the measurements are described in terms of types rather than in terms of quantities. Thus, position is not a single type with many values. Position itself denotes as many types as there are positions. When position and other physical properties are described as types, then the physical properties of the symbol denote meanings. Now, the random sequence of events can be explained because this order represents a sequence of symbols which encode information. A new kind of order now emerges. This order is not a new experience, but it is a new (and more complete) explanation of the same experience. If this explanation works, then we can view reality as a representation of meaning rather than as meaningless things.

The semantic view aids in the resolution of interpretive problems in quantum theory such as the apparent conflict between atomic and macroscopic realities, which has hindered an understanding of the randomness underlying quantum phenomena. The randomness is apparent and not real. It, in fact, is caused by the current attempts in science to describe symbols of meaning as classical particles. Like the meaning of a book can only be probabilistically described by measuring word frequencies, similarly, the nature of reality is incompletely comprehended by measuring physical facts.

Mind-Body Dualism in Biology

The semantic view resolves many contentious issues about mind and matter in current science. Mind-body dualism was introduced in science by Descartes, who insisted that matter is *res extensa* while the mind is *res cogitans*, or substances that are extended and that think, respectively. Quantum theory shows that *res extensa* can itself *represent* meanings although *res extensa* is not the *experience* of meanings. The mind and body are therefore not totally different because they are related as the symbols and the experience of meaning. They are also not identical because the symbol itself is not experience. Symbols are created when abstract ideas are refined by adding more information. A physical object is therefore the development of an abstract idea, and both the object and the underlying abstractions are semantic. A specific space-time distribution of matter represents a specific type of meaning, which can be understood when space-time is treated semantically. The location of this space and time are different types, and as more information is added, the space-time is divided into smaller and smaller parts. There is, of course, a limit to this division, represented by Planck's constant.

Many problems in science can be solved by this view:

- It explains the so-called random events in physics and biology based on a type of logical-mathematical cause that may not always be perceived by the senses, although it could be perceived by deeper forms of perception such as the mind.

- It shows how mind and matter can be different and yet interact. The mind is an abstract idea and matter is a contingent idea. Their interaction is simply the interaction between two types of ideas. When causality is modeled as the interaction between meanings, the interaction between mind and matter is unproblematic. Accordingly, the laws of matter and change in the universe are the laws governing the evolution of meanings.

- It addresses problems associated with the thesis of mind-body identity; it shows that although everything in the mind can be encoded in matter, the mind never reduces to matter because meaning can logically exist without a physical encoding (meanings would now exist as abstract instead of contingent ideas).

- It shows why the functional view is superior to the material view. In the functional view, the ensemble is real and its parts are defined in relation to the whole. In the material view, the parts are independent of any aggregation. Quantum theory shows that the material description pertains to classical physics, and it is wrong. The functional view is consistent with quantum theory and therefore correct, although the present quantum theory would have to be reinterpreted as a theory of functions. Biologists habitually use the functional view while claiming that it is equivalent to the material description. This premise is false. The materialist view can be constructed from the functional view by discarding meaning but not the other way around.

The semantic view changes the relation between reality and experience. In current science, everything empirical is also physical. In the semantic view, the definition of empiricism is broadened to include other kinds of perception. Now, even abstract ideas which cannot be seen, tasted, touched, smelt, or heard, can still exist and their existence can be detected *indirectly* through their effects   on things that can be seen, tasted, touched, smelt, or heard, and *directly* through deeper forms of perception. If the abstract ideas have not been perceived and they haven't been incorporated in science, then the effects caused by such ideas would never be understood and the theories of nature would always be incomplete. If, however, such deeper forms of reality are perceived, not only would they help the explanation of current incompleteness, but also assist in the formation of newer forms of causal interactions with the world.

So far, science has produced theories drawn from ideas about objects, where objects are supposed to be a priori real material sources of causality. This causality, as we have discussed, is incomplete and scientific theories are forced to acknowledge the existence of probabilities, indeterminism, and randomness. In many areas (which include quantum theory but are not limited to it[19]) science has also been able to prove that we cannot provide a better *material* explanation of experience (Bell's Theorem is an example of such a proof). In other words, we will not discover a new type of experience that corresponds to a new type of material object which will bridge the incompleteness of current theories and will explain the observations better than current theories[20]. Where such proofs have been produced, we can conclude that experience is already complete, and we are not missing something in experience that we will discover at a later point through a 'deeper' observation of reality which will help us form a better picture of material reality.

The incompleteness is therefore a fact about the explanation of that experience, and not about the experience itself. If we suppose that the empirical world[21] is also the physical world, we must forever live with the limitations in science, since better explanations cannot be created. If, however, we can acknowledge that everything empirical is not necessarily physical, then we can construct non-physical *theories*. In the specific case described above, physics needs a theory of meaning—which can be understood logically and mathematically—although the meaning itself does have a material origin.

Mind-body dualism is the dualism between meaning and sensible objects. Meanings can exist in objects but meanings are not the sensible objects. Rather, meanings are encoded in objects in order to *communicate* them to other minds. By this encoding, the meaning in some mind becomes publicly accessible to other minds. Other minds can physically observe these meaning-tokens and science can build theories about them. However, theories that study the tokens without reference to meanings will be incomplete. These theories can be completed only by recognizing that the token was created by the meaning that existed prior to and outside of the space-time token. Tokenizing the meaning is useful to communicate our ideas, and we have a great need to communicate them. But the ideas themselves can exist without communication. The public existence of material objects indicates the existence of meaning in the creator's mind. Matter therefore points towards the mind's existence.

Science has historically discarded questions of mind. It sees objects as things-in-themselves rather than tokens of meaning. The notion that matter tokenizes the mind's meanings preserves the empirical and logical nature of science, but shifts its ideological basis from matter to mind. That is, material objects are created when some mind encodes meanings into matter. This encoding creates objects in the space-time vacuum which other observers can see.

The specific significance of this idea in the context of biology is that there is evolution but the laws of that evolution are not based upon random mutations. The laws of evolution are rather about the evolution of meanings. Of course, these laws are not limited to the evolution of biological species. Rather, the evolution of ideas and ideologies, the rise and fall of civilizations, cultures and societies, the ups and downs in the economy all follow the same laws. While evolutionists may apply the notion of mutation to the evolution of ideologies, civilizations, and economies, there is a better explanation for these phenomena outside the evolutionary approach.

This explanation requires the idea that minds and meanings are evolving according to different laws than used by current science. These different types of laws can be seen in a variety of human and natural phenomena—all of which are treated semantically.

To treat objects semantically, science must first treat space-time semantically. And space-time can be treated semantically if space is closed and time is cyclic. Everyday experience illustrates that locations inside a closed space—e.g., a house—are given semantic names, such as kitchen, bedroom, study, bathroom, etc. We are also acquainted with how time—when treated cyclically—is given semantic names such as morning, afternoon, evening, night, days of the week, months, seasons, etc. Quantum theory already incorporates some rudiments of semantic space-time. For instance, ensembles form a closed space and quantum vibrations constitute cyclic time. These ideas need to be formalized into a new type-based theory of matter. A type theory needs a new type-based theory of numbers, which, as I have shown separately[22], resolves contradictions in number theory. There are, hence, several compelling reasons underlying the shift to a viewpoint where reality is material although it cannot be seen, tasted, touched, smelt, or heard.

Modern critiques of evolutionary theory that question the longevity, frequency, and effect of random mutations pale in comparison to the problems that lie in trying to explain the mind in terms of the brain. While the gaps in connecting the DNA of species and fossil remains of such species suffer from significant methodological challenges because much of the data being explained pertains to events that happened millions of years ago, and has perhaps been lost over time, this isn't quite the case for the brain-mind situation, which is available for observation right now. The fundamental problem regarding the mind is that we perceive the world in terms of *types* while science describes the same world in terms of *quantities.* There is no explanation in science for how quantities can become types[23]. Of course, this does not mean that we are not progressing in the study of the brain, because we are. However, this progress is similar to the study of a book without being aware of the language in which it encodes meanings. Brain studies have found a correlation between the existence of thoughts, emotions, and sensations on the one side, and electrical activity in the brain on the other. We know that testosterone stimulates sex drive, serotonin creates feelings of happiness, and cortisol and epinephrine are related to stress. This correlation between molecules and mental states is not false. But it hardly explains how the molecule *becomes* the emotion. There is a big difference between being serotonin and being happy, and current science doesn't have the theoretical tools to bridge this gap. We can dig deeper in the brain and find more chemicals, but the psychological *effect* of these chemicals can never be derived from the chemicals themselves without a semantic viewpoint. Today, these correlations are formed by relating the brain's electrical activity to the person's mental reports. But these are simply correlations between two events, not an explanation of one based on the other.

The problem of mind-body dualism exists in biology as the dualism between happiness and serotonin. This problem can be resolved if we say that serotonin is the symbol of happiness, within a suitable ensemble of molecules. Outside the ensemble, serotonin can have a different meaning. Therefore, the same molecule in the test tube does not represent happiness, because the meaning is given only within a suitable ensemble, and happiness is encoded by the brain, not in the test tube. The meanings represented by chemicals are therefore not universal properties of chemicals as physical properties are. They are rather context-sensitive properties. Indeed, serotonin may not necessarily be the only molecule that encodes happiness; in other species there can be other kinds of molecules. Nor will serotonin always encode happiness and injecting serotonin will not consistently make everyone happy. Happiness therefore is not reducible to serotonin. The experience of happiness depends on a certain space-time distribution of matter as a whole. In a different distribution, the meaning is different, although the physical state of the molecule in question may be the same across the ensembles.

A materialist can claim that since meanings are encoded in matter, the explanation of meaning must be based on the physical states of matter. This claim is false because physical theories are consistent with many possible matter distributions. Physical theories only require the total energy and matter to be constant. Since there are several possible distributions consistent with the total energy and matter being constant, the physical theories underspecify the matter distribution. Each such matter distribution is a different state of the universe and these states cannot be predicted because there is no way to choose one distribution over another.

Towards a Semantic Science

With such a semantic physical theory of nature, we can foresee that other domains of science will also undergo a dramatic change. For instance, molecules in chemistry will not be seen in terms of bonds and structures, which are but physical descriptions of the molecule. Rather, a molecule will be seen as a semantic *proposition* that combines elementary concepts to create complex meanings. Chemical isomers (i.e. molecules that have the same chemical composition but a different structure) will now correspond to two different sentences that have the same word composition but entirely different meanings. This viewpoint is very valuable to understanding why molecular folding patterns make molecules biologically relevant.

It is well-known, for instance, that biological compounds are active inside the body but inactive outside. This difference is attributed to their different folding patterns. Currently, these folding patterns are physically studied as creating sites for catalyzing chemical reactions. The same folding patterns can, however, be studied as representing different meanings inside and outside the body. The DNA which is today studied as molecular structure can also now be described in terms of types produced by combining elementary concepts. Two individuals who have different DNA can be seen as having two different kinds of mental and intellectual qualities[24].

The same meaning can be encoded in different physical structures, and a given physical structure can have different meanings in different ensembles. This is very important because it will now help us correlate widely different types of DNA, something which would have been impossible earlier due to false positives and false negatives. For instance, the DNA of two species may be physically different but semantically similar. This will help science explain psychological and behavioral similarities across species that apparently have different DNA but seem to have very similar types of behaviors, social customs, and food habits. Indeed, semantics will also help us understand why a small difference in the DNA causes the onset of diseases and other changes. Essentially, just as two isomers have the same chemical composition but different meanings given by their different structures, similarly, a small mutation can denote a large semantic difference and thereby explain the effect of DNA changes on psychology, behaviors, and biological functioning.

The study of meaning will make different theoretical predictions than the study of physical properties in matter, even though meanings may be derived from the observation of physical properties. The more complex the system (or ensemble) the greater is the likelihood of variance between semantic and physical approaches. This is because a small difference in physical properties can cause a large difference in meaning, just as a small change in a sentence can dramatically alter its meaning. Likewise, large differences between physical states may only cause a small difference in meanings, just as two physically different sentences can denote the same meaning. If causality is mediated by meanings, as I have argued, these differences will produce myriad empirical confirmations of the semantic view within chemistry, biochemistry, medicine, and neuroscience.

Probability and Semantics

Modern evolutionary theory relies on the idea of random mutations and, philosophically speaking, there are two ways in which randomness can be viewed. First, we can claim that randomness is epistemic, and it represents our ignorance about the nature of causality. Second, we can argue that randomness is ontological, or that nature itself is random. Charles Darwin seems to have held the epistemic viewpoint regarding chance. In *The Origin of Species* he wrote:

> I have hitherto sometimes spoken as if the variations so com-
> mon and multiform in organic beings under domestication, and
> in a lesser degree in those in a state of nature had been due
> to chance. This, of course, is a wholly incorrect expression, but
> it serves to acknowledge plainly our ignorance of the cause of
> each particular variation.[25]

The epistemic chance may have been scientifically unsatisfying, but in the last 100 years, developments in quantum chemistry have implied random changes to molecules. The dilemmas of chance in biology also exist in quantum theory—are quantum probabilities epistemic or ontological? Does the fact that we cannot provide a better explanation of quantum phenomena mean a shortcoming in our theory or a feature of nature itself? Einstein was amongst the quantum theorists who viewed quantum probabilities epistemically. He believed that one day there will be a better explanation of quantum phenomena. However, with the growing empirical confirmations of quantum theory and the complete lack of alternative approaches, scientists have come to treat quantum probabilities ontologically.

The ontological viewpoint not only violates the founding premise in science that nature is rational and knowable, but also contradicts the observed stability of the macroscopic world, as we have previously seen. Quantum probabilities and their conflict with the classically deterministic world is the foremost unsolved physical problem in science today. It is therefore incorrect to prematurely assume that quantum theory proves that nature allows randomness and chance. It is rather more correct to presume that we still don't have a final theory of nature that reconciles the contradictions between chance and determinism, even at the level of physical theories. If and when such a theory will be formed, the new picture of causality might be dramatically different than how it is currently.

Every generation of scientists have felt that theirs was perhaps the last one where major scientific discoveries were made. It is therefore sometimes very difficult to imagine the shape of things in the future. If nature is indeed semantic, and current science tries to study it as physical properties, then it is possible not just to understand why such a science will be uncertain, probabilistic, and incomplete, but also to formulate the ideas on which a new science can be built. The epistemic treatment of probabilities attributes the shortfall to the present scientific view rather to nature. This can be the beginning of a new thinking in terms of symbols and meanings rather than in terms of physical properties. Symbols are also physical, but the *causality* in symbols is not due to their physical properties. Given that the causal explanation in terms of physical properties is incomplete, it is only rational to approach a new form of explanation that can overcome the incompleteness within science.

Quantum probabilities become epistemic in a semantic viewpoint, helping us reconcile the apparent contradiction between classical and quantum theories. Essentially, the macroscopic world is not deterministic as classical physics implies and the atomic world is not probabilities as quantum theory suggests. Both worlds are semantic. Quantum semantics implies there is a limit to the divisibility of concepts and actions; that there is a limit to the smallest concept or action in nature. These 'atomic' ideas and actions can be used to build macroscopic objects and changes, and if we discard the meaning in these objects then it is possible to construct *models* of semantic objects that appear to be deterministically driven by forces. These models are, however, not 'real' if nature is symbols. To overcome contradictions and incompleteness, we will need to move to newer models to provide a more complete account of nature.

That would also explain the dilemmas in current science. First, it will explain why there cannot be a better account of quantum phenomena using physical properties, and thereby justify the probabilistic quantum theory as a final but incomplete theory that uses physical properties. Second, it will show that there can be better theories that are not probabilistic, but these theories describe nature in terms of meanings rather than physical properties. Such a physical theory requires a new mathematics in which the categorical confusion between names, things, and concepts (besides other categories) can be avoided. The quantum formalism shows a path in which there are multiple ways to 'measure' nature, which correspond to different categories. These categories correspond to different kinds of 'senses' that interpret the same symbol alternately as a name, concept, algorithm, problem, etc. There is hence a unique reality although it can be interpreted or modeled in various ways. These models are mutually exclusive in the sense that they correspond to different methods of observation. But they are also simultaneously available to each observer. These categories exist in ordinary language because this language describes experiences.

The meaning of 'meaning' is therefore not one thing. Meanings span a spectrum of ideas ranging from concepts (common nouns), names (proper nouns), algorithms (designs), problems (goals), etc. That these categories represent different classes of meanings also helps deconstruct the 'mind' into faculties that measure different categories. To progress, science needs not just an understanding of how matter encodes symbols, but also of the various types of meanings that can be encoded in symbols. The former is reality and the latter is its observation. Like current science postulates the existence of different kinds of properties in matter, a semantic science will postulate the existence of various types of meanings in the same symbol. Each such type of meaning corresponds to a different faculty in the observer by which we see the world in different ways. To progress in science, scientists will have to find the various ways in which the world is known through observation. Current science is based on observation through the senses. This science tells us about the physical properties in things, but not their meanings. Successive sciences can also explore different kinds of symbol meanings.

Like the present science based on sensations is an incomplete account of nature, future sciences based on a single category of meaning will also be incomplete. A complete science requires the incorporation of every category in ordinary language, including sensations, concepts, names, algorithms, problems, and so forth. Without incorporating all these categories, the predictions of the new science will be incomplete. If we view that description as a final one, we will conclude that there is some randomness in nature. That is a point at which a new category must be added. I believe that the categories in ordinary language and observers are finite. A semantic science therefore reaches an end when all the ways in which we can know and describe a symbol have been incorporated in science.

The conclusion from this shift is that incompleteness (currently seen as randomness) is a problem about not incorporating all the ways in which a symbol can be described. Currently, science describes symbols as physical properties, ignoring their meanings. However, within meaning itself, there are many categorical distinctions. When one kind of meaning is described in terms of another, it leads to incompleteness which the scientist perceives as randomness. If you believe that the theory is final, then that randomness becomes ontological. If, however, you treat your theory as incomplete until it has captured all possible categorical ways in which the world can be known, then the incompleteness is epistemic.

In current biology, randomness is ontological and it forms the basis of evolutionary biology. This randomness can be epistemic in a semantic view. Now, things that seemed random earlier will have a causal explanation based on semantic interactions. These semantic interactions can be between a living being's mind and their body, they can be interactions within a living being's body because molecules and DNA will be treated as propositions of meanings, and finally, there can also be semantic interactions between the living being's body and its environment. All these interactions involve atoms and molecules, but the interaction itself cannot be causally explained based on the physical properties of atoms and molecules.

The Role of the Mind in Science

Biologists often suppose that because the living being's body is made up of molecules all explanations of the body must be based on physical properties. This is based on the premise that the world is fully comprehensible by observation with the five senses (sight, taste, touch, smell, and sound). Modern science is based on the idea that physical properties exhaust everything that can be said about nature. These physical properties are obtained by objectivizing sensations, that is, by finding standard instruments which represent units of such properties. Of course, science cannot work with sensations alone. To coordinate sensations into theories, we require concepts. To formulate concepts, we must aggregate and analyze a variety of different types of sensations. Senses themselves cannot aggregate the sensations or analyze and order them. This job is done by the mind and intellect in the scientist. Thus, while science ultimately depends on the existence of mind and intellect, the role of the mind or the intellect has been minimized by assuming that these do not add anything new to our picture about nature beyond what could have been gathered by sensual observations alone.

Does the mind just organize sensations or does it also *perceive* other things that the senses cannot? When we see a laughing person we conclude that they are happy. We might see knitted eyebrows which imply that the person is thinking. We might observe a person's tone of voice and conclude that they are anxious or depressed. We might experience the sensation of touch and conclude the touch is one of friendliness, lust, or affection. These conclusions are not based upon actual experience of happiness, confusion, thought, depression, friendliness, lust, or love in the person who we see or who touches us. Indeed, we might be mistaken in arriving at these conclusions, quite like we might be hallucinating or under a perceptual illusion. But often we are right. The conclusions are purely derived from the observation of the world. And yet, they are not obtainable simply by the five senses of taste, smell, color, sound, and touch.

The observer's mind and intellect do not simply organize the sensations into abstract concepts. They also add meanings to these sensations. Modern science has assumed that these additions are subjective and not relevant to the study of material objects. But they seem very pertinent when the material object in question is a fellow human being or animal. They are also pertinent when material objects encode meanings—such as in the case of books, music, paintings, and science. Indeed, most readers draw conclusions about the author's intentions from the book itself. For instance, many of you would have already drawn some conclusions about my intentions behind writing this book. Whether you are right about those conclusions or not is beside the point. The fact is that by seeing the tokens in this book, you can not only find conceptual meanings but also understand the writer's emotional state and intentions from it.

The analysis of sensations by the mind and intellect therefore are not limited to the study of their physical properties. Rather, we can derive various categorical meanings from the sensations themselves. How do we do it? It seems that we attribute new meanings to the sensations based upon the space-time context provided by other sensations. In other words, the mind and intellect analyze sensations to see what they *represent* besides what they *are.* Curly lips can denote happiness, disgust, confusion, or disrespect. Only a careful analysis, coordination, and correlation of the individual sense-data can lead us to one of the several possible meanings. In science, the correlation and coordination of sense-data attributes causal relations between the sense-data, but discards the meanings we commonly associate with the sense-data. This is not just a limited view of the mind but it also becomes limited when additional types of meanings could be derived from the same sensations. The need in science is to widen the role the mind plays to incorporate other forms of coordination and correlation than those currently practiced. That is, the mind will not just perceive the order amongst events and think of causal connections, but also associate meanings with this order. Like new meanings emerge from the coordination of sensations, which would not be possible in the individual sensation, similarly, new physical properties can be perceived by giving reality to the semantic coordination and correlation activities of the mind.

Humans are distinct from other animals with regard to their mental capabilities to coordinate and correlate. Thus, we can see more meanings that animals. There is a wider repertoire of well-developed semantic categories in humans than in animals. If knowledge of the world depended only on sensations, then other animals that have a better sense of sight, sound, or smell would have more advanced theories about nature than us. They would then have more symbolism, pictures, music, and science than humans. That doesn't seem to be the case. This difference can only be attributed to a better developed mind in humans than animals. The mind helps us discover new properties that exist in collections rather than in the individuals. Why would we not attribute these properties back to the collections? Why would we claim that collecting is a mental activity while things are essentially individual? Why would we believe that the coordination and correlation by the mind are useful as long as they don't attribute new properties to the collections? This seems more of a metaphysical bias about the nature of reality rather than anything logical or scientifically justified.

Science stands at a juncture where problems of incompleteness, indeterminism, uncertainty, inconsistency, and randomness, abound. A methodological shift that widens the mind's role from merely distinguishing and ordering sense-data to also giving it meaning can change the current predicament. The role played by the mind lies in solving the incompleteness by supposing that material objects are symbols whose properties are their meanings.

This viewpoint solves the problems of meaning *perception* but not the problem of meaning *creation*. The perception of meaning follows the creation of meaning, and is therefore a more important question. If meaning arises from the interrelation between parts in a collection, then the problem of meaning creation is how you create and distribute the parts. We saw this problem previously in atomic theory where alternative distributions of matter subject to total energy remaining constant are consistent with the theory. We also saw this problem repeated in the case of statistical mechanics and general relativity. Atomic theory is incomplete not just due to probabilities but also because it does not predict which distribution of matter is real. Statistical mechanics and relativity theory are similarly indeterministic because they don't predict matter distributions.

If there are many matter distributions for the same total amount of energy, and these distributions represent meanings, then meaning must be the *cause* of the distribution. Now we cannot postulate the existence of *a priori* real particles which are randomly being structured. Rather, choosing an eigenfunction basis or matter distribution in statistical mechanics and general relativity implies the choice of particles and their states. Therefore, to choose a particular distribution there must be a mechanism to construct the particles *and* the distribution *at the same time*. This is impossible in current quantum theory. John von Neumann proposed that the choices of consciousness collapse the wavefunction into a definite measurement, but he did not propose such a mechanism for choosing the eigenfunction basis. If we extend von Neumann's idea to the choice of basis, these will be two kinds of choices: one that selects a distribution and the other that picks a particle in that distribution.

Clearly, the choice of distribution precedes the choice of a particle in that distribution, and the choice of distribution is therefore logically prior. Both these choices are mysterious in the current physical view of nature, but they become tenable in the semantic view. Now a distribution of matter represents some encoded meaning that is decoded during observation. The order of symbols in the measurement is like the order of words in a proposition. However, the same meaning can be expressed through different words and their order. The eigenfunction basis represents a choice of words and the order of events during measurement represents the order of words. Similarly, the indeterminism in statistical mechanics can be attributed to the fact that there is less than maximal information in the system, and the indeterminism in relativity can be understood as the fact that different observers can have different experiences even when the events in the universe are unchanged.

In current physics, we can measure the states and order amongst these states, but we cannot know if some order of states represents the same or different meanings. This is similar to the fact that different words can convey the same meaning and the same words could convey different meanings. Indeed, the words used to encode meaning may not be the words seen during decoding. Thus, we cannot suppose that the individual particles are 'real.' What we can suppose, however, is that the meaning encoded by the ensemble is real and this meaning could be expressed using different words. This implies that meaning is objective but the words by which that meaning is expressed are a product of the measurement. We can know the same meaning through different words, although we might encode the meaning using a specific set of words. The words themselves are not real, and they only correspond to the language or vocabulary being used to encode and decode meanings. The shift in vocabulary represents a semantic coordinate transform that picks a specific vocabulary to describe the encoded meanings.

This clearly tells us that what science has held dear for the last 400 years—namely the idea that matter is real and meanings are in the mind—is false. Meanings are objective, and more real than the things by which these meanings are expressed. Like we can change the words to express the same meaning, atomic theory shows that it is possible to describe reality in many different ways. What science calls objects are created at the point of measurement but they do not exist *a priori*. It is meanings that exist *a priori* and they are converted into words (matter distribution) during measurement.

Mind has a role in creating meanings. The role of the senses is in observing those meanings as sensations or words. The mind also correlates those words to reconstruct the meaning. There are many ways in which we can observe the same meaning—e.g., we can touch, taste, smell, or see the same food, each time describing it using different words (the words of taste are different from words of color, and food when tasted will not have color although it will have taste). What the senses perceive therefore is a *description* of reality and not reality itself because the same reality can be described differently by different kinds of senses. The different descriptions of reality are, however, not arbitrary. They are related through a mathematical transform, when mathematics is defined as the study of types rather than of quantities. Such a theory will produce new predictions. For instance, if we knew how the food tastes, then we will also be able to predict how it looks (something impossible in current science). But, given that we can know and describe reality in many ways, each time using a different sense or combination of senses, the same information will be expressed in different ways.

Materialism is the idea that matter is real and meaning is epiphenomenal. The problems of modern physics, mathematics, computing, and biology point towards a new idea of *mentalism* where the meanings are real and the objects are created to express those meanings. Nature is therefore not *a priori* real objects which move in space-time, as classical physics claims. Rather, the objects that we perceive are created by the observational apparatus during observation. This does not imply that there is no reality. It only implies that the reality is meanings and not objects. The physical world therefore does not exist until we observe it, but that does not mean there is no world out there. The world actually exists as meaning. Perception manifests the objects from that meaning, just like words are manifested to describe a meaning. We can thus know the world in many ways, but not arbitrary ways. To know the reality behind the appearances, science needs to treat the observations semantically where each type of observation will produce meanings.

The Idealist View of Nature

The semantic alternative to evolution is not just a view about living beings having a mind. It is rather about material objects being representations of ideas. In Greek times, Plato postulated that the world of everyday objects is a reflection of a world of pure ideas. Thus, the tables or chairs in this world were reflections of the pure ideas of a 'table' or 'chair;' the world of ideas had to exist before the world of things could. With the Cartesian separation between mind and body, ideas and things were divided into two different worlds and with the subsequent attempts to reduce ideas to objects, the world of ideas simply ceased to exist. The reduction of ideas to things hinges on the success of the materialist belief that material laws are consistent and complete and a mathematically consistent and complete description of matter *can* be formulated. This, as I have discussed earlier, is impossible and problems of indeterminism, incompleteness, incomputability, and irreversibility abound in current science. There is hence an imminent need to retrace our steps back to the Platonic view about the relation between ideas and things, reinstating the primary role for ideas within nature.

If the reduction of ideas to things is known to fail, can we do the inverse of reducing things to ideas? This, as I have elaborated, is possible but requires a rethink of practically every scientific concept, including numbers, objects, and laws. Numbering is a property of a collection of objects, so numbers cannot exist until objects and collections can exist. The incompleteness of physicalist theories shows that there are many ways to divide a collection into its objective parts, so the collection must be logically prior to the parts. Indeed, we cannot suppose that objects are real *a priori* and their collection creates epiphenomenal meanings and minds. The incompleteness of all theories that use the idea of *a priori* real objects indicates that although there is something real in the world (because otherwise science would be outright false), it is not what we currently think it is. As we saw earlier, the problems of incompleteness indicate that the collection is real, but how it is divided into parts is not determined. By finding new ways of thinking about collections as being logically prior to objects, it is possible to complete science. If collections are logically prior to the objects—and objects are produced by distinguishing and ordering the collections—then the laws of nature must pertain not to the objects themselves but to the acts of dividing and ordering collections to produce these objects.

The Platonic view is also idealist, but it postulates the eternality of all ideas: tables, chairs, and beds are eternal ideas. This is hugely problematic if science has to be constructed from a parsimonious set of ideas. Platonic Idealism therefore does not work for science, and a new way of constructing ideas is required for Idealism to be a scientific idea. This is possible if ideas themselves are created from the choices of consciousness. The universe can begin in a parsimonious set of ideas (the *axioms* in a mathematical theory of ideas), but these ideas have to be mutated and combined to create more ideas. The mutation and combination of ideas can be logically described as *possibilities* of idea construction in a mathematical theory but the *act* of combining and mutating must remain a choice in such a theory. That is, the mathematical theory will describe everything that is possible in nature, but not which things will be real and how. Accordingly, the laws of nature would be the consequences of these choices: they will dictate the effects and reactions of choices.

Both Platonic Idealism and scientific materialism are unsatisfactory theories of nature; idealism is unsatisfactory because it is not parsimonious and science is unsatisfactory because even if it is parsimonious it is incomplete. The incompleteness in science can be traced to the fact that we are trying to describe a world of meanings produced from choices in terms of meaningless things governed by deterministic theories. The ideological evolution in the Western world has vacillated between idealism and materialism for the last two millennia, not coming to a satisfactory conclusion although reinvigorating the debate each time a shift appeared to occur. I believe that this vacillation in ideology is primarily due to the fact that we haven't recognized the central role for consciousness which produces ideas, which are then responsible for producing objects.

Therefore, the sense in which I use the term 'idealism' differs from the sense in which it has been used in the past in the Western world. My use of 'idealism' indicates the existence of a consciousness which produces ideas, which in turn create objects. The acts of idea production itself cannot be understood within science, although the consequences of the idea production can be described. Science as the exercise of describing *if-then* scenarios can still be formalized, although the "*if*" (cause) is now traced to a choice.

Implications for Evolution

One of the fundamental ideas underlying evolution is that nature produces fundamental sub-atomic particles, but the production of atoms and molecules from these particles is not pre-determined. Mutations of molecules therefore produce new molecules. These molecules combine to form cells, which create organisms. There is inherent randomness in nature which allows many possible alternative realities. To choose between these realities, biologists postulate the idea of natural selection, which determines *compatibility* between different parts of a given reality. Quantum theory shows that this is clearly not enough because there are infinitely many realities that are mutually compatible. How do we single out one reality from these infinitely many possibilities? The answer must again be randomness, unless we find a new causal mechanism. In that sense, natural selection does not add anything new to what physical theories already tell us. Natural selection only says that parts of reality must be semantically compatible (e.g., give-and-take relationships of food, shelter, recreation, sex, etc.). Given that there are infinitely many ways to construct these mutually compatible matter distributions, natural selection does not solve the randomness problem.

We also saw above that a similar conclusion is implied by other physical theories such as statistical mechanics and general relativity which allow infinitely many matter distributions which are all mutually compatible, and consistent with the laws of conservation. There is no law of nature that singles out a matter distribution.

The only solution to this problem is to treat the randomness as being epistemic rather than ontological. The epistemic randomness can be bridged by treating matter semantically. But, now, meaning precedes material objects. Meanings are not byproducts of mutating atomic objects into molecules, cells, and organisms. Rather, objects themselves are byproducts of embedding meaning in matter. Indeed, objects are produced during the observation of meaning. To create these objects, there must be meaning prior to matter. And to produce the meaning there must be a mind prior to the objects.

The evolutionary notion that complexity in matter produces a mind over millions of years of evolution contradicts what mathematics and physics tell us. Mathematics tells us that there can never be a consistent and complete theory if there is meaning in matter and we describe this meaning as a byproduct of matter. Physics tells us that if reality is material then there are infinitely many mutually consistent possibilities (matter distributions) to choose from. The information gap between what science predicts and what we actually observe is not large or huge; it is infinite. There is simply no physical mechanism that we can postulate that will overcome this incompleteness, because such a postulate will lead to inconsistency (Gödel's Incompleteness in mathematics and Bell's Theorem in physics). The evolutionary idea that we can postulate additional physical mechanisms that will overcome the gaps in physics and mathematics to create certainty and meaning in matter are therefore contradictory to what physics and mathematics assert.

The path forward is to postulate that meanings are real and objects represent meanings. When information is added to abstract ideas, objects are created. This additional information can be supplied during state preparation or even during observation. In other words, the observer has a key role in defining and observing reality. If objects were *a priori* real, there would be only one way to define and observe nature. The many possible matter distributions of nature (given a total amount of energy) indicate that nature can be defined in many ways. And, when nature has been described abstractly, its state can be refined in many ways by adding information. Therefore, reality cannot be the objects we observe. We must suppose that reality exists as meanings which are expressed as symbols, and we perceive the symbols as sensations of things. There is a reality, but that reality is not things; it is objective meanings.

This view of nature is the only consistent way forward to reconcile the problems in mathematics, physics, and computing. And yet, it also undermines the ideas underlying evolution, without denying evolution. Now, evolution is not about random mutation and natural selection. Evolution rather pertains to meanings. The rise and fall of civilizations, nation states, cultures, ideologies, and ideas are all examples of this. Biological evolution is a longer timespan change than the other kinds mentioned above, but it can be modeled on the same ideas. This view of evolution however needs an entirely different understanding of nature than in current science. It necessitates primacy of meanings rather than of things. It also implies that the Cartesian separation between mind and matter, and the subsequent attempts in science to undermine the role the mind plays in matter, are both false. We cannot study objects independent of the mind because the objects don't exist unless someone observes them. What exists is meanings which must have been created by prior minds to communicate with other minds.

The evolutionary idea that minds arrive late in evolution is false if objects are created at the point of observation, simply expressing the information that was prior encoded by the mind. The mind is not a *post-hoc* addition to some material configuration. The mind is the precondition for any kind of object to exist in nature.

The biggest challenge for evolutionism is not creationism. The biggest challenge is indeterminism, incompleteness, and uncertainty problems in mathematics, computing and physics. To the extent that we know these problems cannot be solved within the current materialistic theories, science has hit a logically impregnable brick wall. The laws of random mutation and natural selection that biologists postulate must be within the laws that physics and mathematics permit. And these theories are incomplete. There is nothing in current physical theories that biology can rely on to complete the current incompleteness. Further evolutionary ideas that deny a logically prior existence of meaning and mind are not per se denied by mathematics and physics but it is known that they cannot fit current mathematics and physics without creating logical contradictions. There is, hence, never a point in which the current materialist outlook will explain the existence of concepts and meanings in matter. It is logically impossible for science to explain its own existence without invoking a separate mind and meaning outside of matter.

Given these problems, I consider it highly unlikely that evolutionary theory is true. That doesn't deny evolution. It only says that the *mechanisms* and *postulates* of the current evolutionary theory are false. These mechanisms include the idea of random mutation and natural selection. There are mutations in nature, but they should be explained on the basis of change in meaning; they are not random. Natural selection is a semantic compatibility between concepts which require a mutation to take place prior, which is in turn governed by changes to meaning. These two changes to evolution require the mind and meaning to be real *before* objects are real.

4

An Alternative Evolutionary Theory

In the light what is known about the radiocarbon method and the way it is used, it is truly astonishing that many authors will cite agreeable determinations as a "proof" for their beliefs. The implications of pervasive contamination and ancient variations in carbon-14 levels are steadfastly ignored by those who based their argument upon the dates. The radiocarbon method is still not capable of yielding accurate and reliable results. There are gross discrepancies, the chronology is uneven and relative, and the accepted dates are actually selected dates.

—Robert E. Lee

What Is a Species?

The theory of evolution started with the problem of explaining the diversity of species. It was not difficult to see the similarities between physical traits of many living beings, which led to the original idea of a species. But, unlike Mendeleev who classified the chemical elements into the Periodic Table, biologists were not satisfied with just a classification. Biologists also wanted to explain the *origin* of all species from a single proto-cell, the cell from molecules, and so on.

The semantic view changes our notion of a species. A species is now not a type of body, DNA, or bag of molecules. A species is rather a type of mind. The bodily shape, the chemicals, and DNA etc. are how we sensually perceive the meaning originating in the mind. In Greek times, Plato and Aristotle postulated that ordinary objects such as tables and chairs have the idea of chair and table prior to them. It is possible to think about biology in a similar manner if we suppose that the body of a dog or a cat represents the idea of a dog or cat. While in Platonism the ideas were in another world, it is possible also to think that the ideas are produced by the mind.

The body of an animal can be seen as a representation of the idea of that animal. In Cartesian dualism, the mind captures ideas about the world and represents the world. But if what that mind perceives is true, then the ideas must have prior been produced by some mind and represented in matter. The mind can observe the world as ideas only if the world was previously created from ideas.

In the case of a material object—such as a table or a chair—the mind that produces the representation of the idea has a different body from the representation itself, although the objectified meaning bears the imprint of the creator's mind. In the case of a living body, much of the meaning that we associate with the living body originates from the mind that we generally think 'lives inside' the body. The mind is actually not inside the body. The mind is an abstract idea which cannot be perceived because the perceptual senses are themselves more refined and contingent than the mind. The mind is the abstract idea from which the contingent ideas (such as senses and the body) are elaborated. In that specific sense, the mind is the source of information which when refined and detailed becomes the objective symbols of that mind. The reverse process occurs during perception: this detailed meaning is abstracted as sensations by measuring the 'macroscopic' aspects of the objects.

Science studies these sensations and attributes them to physical objects. In classical physics, these objects were particles and waves and they were supposed to be materially real and existent in space-time. Quantum theory, however, shows that sensations are produced at the point of observation, and which sensations are produced depends on the choice of eigenfunction basis. Therefore, it is no longer possible to suppose that there are *a priori* real objects which different observers will see in the same way. It is rather truer to suppose that there is a reality from which sensations are produced when the observer's senses interact with the reality. The choices of observation let us perceive different kinds of sensations from the same reality. Therefore, sensations cannot directly correspond to material objects as they exist prior to observation. Rather, we must postulate that these sensations are created from a reality that exists but is different from the sensations. The sensations—e.g., position—depend on the choice of observation, but the reality prior to observation is not at the same position as during observation.

The reality that lies behind the perceptions is the meanings that have been produced as information from some mind prior. The types we see in the bodies are actually produced by different types of minds. Since the same meaning can be expressed in different ways, the type of mind is underdetermined by the physical structure of the body. Two widely different structures can represent the same mind and two similar structures may have quite different minds. This means that there can be a human body with a dog-like or tree-like mind, and there can be a human-like mind with a dog or tree body. Accordingly, the behavior and characteristics of that individual are not determined by the type of body but by the type of mind. The species is not the type of body but actually the type of mind.

Current notions about species are governed by the physical structure of the body. This presents great difficulties in explaining widely different behaviors emerging from the same genetic pool. In fact, the relation between DNA and psychological characteristics itself is very contentious. Is there a gene for artistic creativity? Can we know the perfect theory of reality by creating a physicist with a perfect mathematical physics gene? Why can't we create the perfect president of a country by adjusting the politics gene in the DNA? Or maybe the richest man on Earth by modifying the richness gene?

Intuitively, we know that the psychology and personality of a person involve much more than molecules. They require patterns of thinking that are influenced by our past conscious and unconscious experience, circumstances to nurture native propensities, opportunities to use the skills, and often cataclysmic on-going movements to take advantage of social trends. It is in the confluence of many factors that a person's psychological behavior is given. These factors can only be studied semantically when we explicitly define a science that deals in collections rather than individual objects. The genetic material in the DNA can certainly encode psychological propensities, but *how* a molecule influences the psychology cannot be explained unless the molecule itself is defined in terms of meaning.

As we saw with Gödel's incompleteness, it is possible to define objects using types, but it is not possible to define types using objects. The notion, therefore, that species are physical structures which then acquire mental traits is flawed. The converse notion that species are minds which then acquire bodies helps explain how similarities in minds can lead to similarities in physical structures. The similarity between species therefore cannot be understood by the analysis of the physical structures alone—DNA, behavior patterns, physical traits, etc. As we saw previously, any object can perform different functions in different collections. If we only look at the object, and reduce the function to the object, the description would be incomplete. That incompleteness forces us to recognize both objects and functions. Now, to avoid the problem of inconsistency, objects must be produced from functions, rather than functions from objects. That means that the body's origin must be described based on the mind, which encompasses the semantic traits of the living being.

The similarity between species therefore cannot be understood based on the physical traits alone, because these physical traits underdetermine behaviors. However, it can be understood based on the mind. A species is the type of mind that causes the body. The evolution of the body and the appearances and disappearances of various kinds of bodies are essentially the evolution of the mind and the appearances and disappearances of various types of minds. The laws governing the evolution of biological species therefore aren't laws of physical structure; they must be laws of the evolution of the mind. Just as ideas spread and withdraw, as civilizations and cultures rise and fall, as economies and societies emerge and collapse, the biological world is also governed by cyclic ecosystem changes.

This evolutionary theory is not the random mutation of parts followed by natural selection amongst the parts. It is rather the evolution of the wholes followed by natural selection of the parts that do not fit into the new whole. That is, there are large scale cataclysmic changes which dictate and determine the small scale changes; things that don't fit into the large-scale scheme of things are naturally eliminated. To formulate theories about large-scale changes, science needs to think of nature semantically rather than physically.

The Problem of History

A fundamental premise in current physical theories is that all object states are how they are in themselves; these object states could not pertain to a description of other objects. A second fundamental premise in physical theories is that all object states denote the current state of affairs; an object has no memory of the past. These two ideas are fundamentally violated when we look at fossil records. First, we suppose that a fossil record *refers* to something that is no longer apparent—i.e. the physical form of some animal body. Second, since we cannot find that animal anymore, we suppose that the fossil indicates the existence of something that lived in the *past*.

A fossil is a symbol of the animal which existed in the past. However, to arrive at this conclusion, we must attribute to the fossil two basic properties which do not exist in current physical theories: (a) intentionality, and (b) historicity. Material objects—in current science—do not refer to other objects and they do not describe the past. Therefore, if we tried to understand a fossil purely in terms of current physical theories, we would view it as an *object* rather than as a *symbol* that describes the form of the animal which existed in the past. Of course, this is not a fact unique to fossils. We might as well read a book that physically consists of squiggles, although these squiggles are interpreted to be representations of the facts about other objects, and these facts may have existed in the past. The main difference between a fossil and a book is that we suppose that books are written by humans and fossils were created by nature. However, the intentional and historiographical problems in associating the squiggles in the book to events in the past exist in the case of fossils as well. Quite specifically, we have to view the fossil as a symbol of the past events. If we do not view the fossil as a symbol, then it is just an object that exists in-itself, here and now, and it could not indicate another object especially if that existed in the past. The act of interpreting a fossil as a historical record violates basic assumptions in physical science—namely that things are not symbols. As a symbol, the fossil *points* to another object that exists in the *past.*

The use of fossils to understand the evolution of species, followed by the reduction of species to atoms which do not have intentional and historiographical properties is itself inconsistent. Either matter has intentional and historiographical properties or it does not. If matter has these properties, then they must be present even in living beings. If matter does not have these properties then we cannot use fossils to understand the historical evolution of life.

Of course, I have argued that matter is semantic and does have intentional, contextual, and qualitative properties. Therefore, it is not incorrect to use fossils to study history. But we must recognize that this ability involves a different view of matter than used in science. If fossils are just things (like all things) then we cannot treat them as symbols of knowledge about the past and we cannot draw inferences about the past using these objects. But if fossils are symbols (like all symbols) then matter has semantic properties in itself.

The basic problem of history is that it is inconsistent with physical theories. We use facts that we can obtain now to make inferences about the past, although we don't have a deterministic predictive theory that can accurately connect the past to the present. For instance, if biological evolution was governed by a deterministic theory (like Newton's laws of motion) then given the present conditions we could compute the earlier conditions. But we don't have a deterministic theory of evolution. In fact, evolutionary theory itself claims that biological evolution is not deterministic; it is produced from random chance events followed by natural selection. So, evolutionary theory eliminates a deterministic alternative by supposing that the mechanism for evolution is random mutations. It follows that in this randomness viewpoint, we can never make a *physical* connection between the past and the present (such a connection is possible in a deterministic theory such as Newton's mechanics). Then how can we connect fossil states to the states of past beings?

This connection is possible only if we treat fossils as symbols of information of the past. Furthermore, this information must be *objective*, because otherwise, the interpretation that some fossil describes the living beings in the past would only be in our minds!

There are clearly two ways in which we can interpret the picture in Figure 17. First, we can view it as a rock which has some curious shapes carved into it; these shapes are the current state of the rock and have no other significance. Second, we can view the rock as a fossil which was created by an animal's body in the past. How do we arrive at the second conclusion? We obviously treat the physical states of the rock as telling us about the animal in the past. If we truly treated the fossil as a rock—i.e. a thing that only tells us about itself and not about other things—we could not conclude this way.

Figure-17 How Do Fossils Indicate the Past?

When we read books, we commonly suppose that the book is physical and the meanings are in our minds. In other words, meanings are our *interpretations* of some facts in the world. We also commonly suppose that there are many ways to interpret a book, and thereby derive many different meanings from the same physical structure. All these meanings are imaginary because the meanings don't themselves exist in the book (which is nothing but physical states). If this view were to be extended to fossils, then the idea that a fossil's physical states are descriptions of the states of the animals that existed in the past would also be in our minds. The conclusion that there were indeed animals in the past which had the same physical form as the fossil indicates would just be our interpretation of the fossil. Like a book can be interpreted in many ways to create many meanings and all these meanings rest in our minds, similarly, the fossil could be interpreted in many ways and these interpretations would all be imaginary. Of course, I am, by no means suggesting that this is indeed a correct way of looking at fossils. I am, however, saying that if fossils are treated solely as physical objects, then this conclusion is inevitable. The only way to treat a fossil as a historical record is if we treat matter as a symbol of information.

Does the Past Exist in the Present?

From a physical theory standpoint, we cannot *observe* the past; everything that exists right now and can be observed only denotes the present state of affairs. If we have a deterministic theory, then we can *infer* the past from the present, by tracing back the previous states from which the present state was caused deterministically. But if we don't have such a predictive theory then we cannot infer the past either. All observations of reality must now be understood as the present state of reality and not of the past. For instance, we don't look at an atom or molecule and conclude that it is a historical imprint of the past, and its position and momentum can tell us about the states of other objects, let alone objects that existed in the past. However, we do look at fossils to arrive at that conclusion.

There are thus at least two ways to understand the physical states of a fossil. First, we can treat them as the current physical states of those objects. Second, we can treat them as symbols of the past states of other objects. Biologists and evolutionists routinely draw the second conclusion. While the physicist does not treat the position and momentum of the atoms and molecules as a historical record, the biologist does treat the collection of these particles as a reflection of history. The contradiction should be palpable to everyone: quantum theory does not even describe matter as information about other objects, let alone objects that existed in the past. So, ideally, we must conclude that the fossil is just a physical state. But we don't. We treat the fossil as a symbolic record of the past events.

This raises some fundamental questions: If the fossil can denote the past, then should matter at fundamental levels must also be a representation of the history of the past? If we deny this possibility, then when the macroscopic world is constructed from the atomic world, the history would be automagically created. We routinely suppose that this is indeed what happens. For instance, many materialists argue that the mind is an epiphenomenon of physical interactions; or that the mind is like the fluidity of water. There is no mind in the atoms, and hence nothing is lost if we reduce the mind to the atoms. When this idea is extended to fossils, it would mean that the atomic world does not have any history although the macroscopic world appears to have an historical imprint of the past.

By the logic that we treat the mind as an epiphenomenon of atoms, we would treat the historical impression as an epiphenomenon as well. In other words, the existence of history is an illusion. We are interpreting the molecules in the fossil as indications of the past, but the history is not in fact there. Just as the mind is an illusion of atomic interactions, the history in the fossil is also illusion. If the belief that mental intentionality is false, the belief that fossils can tell us about the past must be false as well. Now we cannot use the fossil to understand the history of biological evolution. Again, I'm not at all insisting that we cannot use the fossils to study history. I'm only insisting that our ability to use the fossils to speak about the past is inconsistent with physical theories. If we have to use the fossils to study the past in biology, we would also have to revise our current notions about physical reality. We would have to allow the possibility that matter can encode information, which can in turn point to other objects, be they objects in the present or in the past.

Does the past exist in the present? Certainly it does, but only as information about the past, not as the past itself. To allow this view, we have to distinguish between the past and information about the past. To derive the information about the past from the present, we need to interpret the present semantically (i.e. as a symbol of the past). That in turn implies that the information must be real, because if it wasn't then the idea that fossils indicate the past would be an imagination of our minds. Without the semantic interpretation, we would see a rock, although not a fossil. However, if the information is real then matter has the ability to hold information and it is objective. In such a scenario, the body of the animal (which is another material object) is also an encoding of information, quite like the fossil is an encoding of information. These two objects will point to the same information—i.e. the mind that created the body.

The difference between these two encodings is that the animal's body was developed on a 'living' mind and the fossil is a historical imprint of that living mind. In short, fossils represent not the archeology of the bodies, but an archeology of the minds. Like a book that records the ideas of the current time, the fossil recorded the idea that existed at a certain time. The fact that we don't find such bodies only means that these ideas don't exist currently—there are no minds that create such ideas right now. However, the ideas themselves are eternal; the fossil encodes these ideas quite like a book might encode or represent the thinking of its times.

Again, does the past exist in the present? Yes, the past always exists, if the past is viewed semantically as ideas; all ideas always exist as a possibility. However, they may not be realized into physical representations. Mathematical Platonists accept this view about mathematics. For instance, a Platonist would say that all mathematical theories are eternal as ideas; they have always existed, although they manifest into our minds occasionally. When they do manifest, we can also write about them and make them accessible to other minds. A mathematical theory is therefore not our invention; the theory is rather a 'discovery.' Astute physicists would also adopt this view and claim that we are discovering the nature of reality not inventing it, although the theory we have clearly did not exist in the past. And yet, a technologist would claim that they invented the computer or the airplane. This creates a somewhat double standard about ideas in modern thinking. When convenient, we suppose that ideas are always possible although they become real only at some times. When inconvenient, we claim that we invented the ideas.

This double standard about the relation between ideas and things can only be resolved by a shift in science in which all reality exists as ideas, which are then embodied into matter when the ideas manifest in the mind. The role of science is now to explain the manifestation of the ideas in the mind. Why are some ideas manifest or unmanifest? In the context of biology, this means the ability to explain how some species or forms are created and annihilated.

The past, present, and future are always present if reality is described as ideas. The difference between past, present, and future is not one of idea creation but one of idea manifestation. When the body is a representation of these ideas, then the evolution of bodies (or species) must also follow the laws of idea manifestation. As we noted earlier, in an idea-like universe, space and time are hierarchical and closed. That in turn implies that ideas must be manifested and annihilated cyclically. The ideas that exist in some part of space may be annihilated in that part of space only to be created later in another part of space. Also, the same place will produce and destroy the same ideas over and over again, consequently creating and destroying the same types of living bodies over and over again.

Towards a Mathematical Theory

Classical physics modeled the universe as a network of interactions between *a priori* real objects. In Newton's gravitational theory, for instance, the universe is a collection of particles which exert a gravitational force on all other particles in the universe. This interaction between particles could be depicted through a network diagram in which each object has an interaction (shown through a connecting line in Figure 18) with every other object. The big issue with this picture of nature is that it assumes a set of *a priori* real particles; for instance, if the particles are not real, then we cannot draw interaction diagrams. Later developments in science have shown that nature is in fact not a set of *a priori* real particles. Matter is, rather, fungible since the same amount of matter or energy can be divided in different ways to create various sets of distinctive particles.

Quantum theory shows that nature is not conserved in the total number of particles; it is conserved in the total mass, charge, energy, momentum, angular momentum, and spin. Quantum theory indicates that there isn't a fixed network diagram that we can call 'reality,' although there is some reality which is converted into a real world of objects during observation. Indeed, there are many network diagrams that could be constructed by moving matter and energy to create different particles. Since there is always a single reality upon observation, how one amongst these infinite possibilities is selected to create reality remains a mystery in quantum theory.

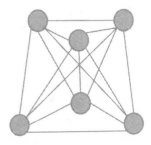 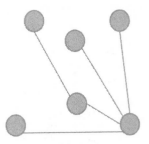

Figure-18 Network and Tree Topologies

The network diagram also creates problems of counting or ordering the objects. For instance, the network nodes can be traversed in many possible ways, each time labeling the nodes with different numbers. All these different ways of counting correspond to different network paths. Of course, we cannot traverse the same node twice in a network (because then we would be counting the same node twice). To prevent repeat traversal, each method of counting must reduce the network to a tree. In a tree, each object is only traversed once and hence only counted once. However, even when we convert the network to a tree, we still have many possible ways of labeling these objects to numbers. Each such way of counting objects can begin with one particular node in a tree and label that as the first node. Accordingly, other nodes will be called second, third, fourth, etc. In physical theories, each method of counting represents a coordinate transform, and physicists have supposed that nature allows infinitely many ways of counting the nodes in a network.

The assumption in physics is that coordinate transforms simply label objects with different numbers, and these labels are inconsequential as far the real properties of the objects are concerned. The labels by which we call these objects represent their *names* and any object can be called by any name; the changes to names in different coordinate systems would not affect the behavior of objects. This premise in physics is true when matter is treated as physical properties but becomes false when objects are treated semantically. In a semantic treatment, the names by which we call the objects must also denote their properties, unless the distinction between names and meanings can be instituted in another way. The distinction between names and meanings leads to mind-body divide—it is the observer who calls the world by different names and gives them different meanings, although the world has no name or meaning. The mind-body divide in turn leads to many other problems as previously seen. The ability to give arbitrary names to objects leads to Gödel's incompleteness where an object can have a name opposite to its properties, and when the name is interpreted as the property (as the mind-body divide is collapsed) contradictions follow.

For instance, a barber can be called Mr. Potter and since the distinction between name and meaning is not kept, when the name is interpreted as the meaning, the naming contradicts the meaning.

This problem can only be solved in one way—we must begin counting objects from an object which is also the semantic root of the other objects. In other words, the first object in that tree is also the first meaning from which all other meanings are constructed. This approach to counting prevents a barber from being called by the name Mr. Potter, and thereby avoids logical contradictions.

The locations of objects in space-time are now the *names* by which they are called and these names are consistent with the meanings of those objects. The universe is a hierarchical tree constructed from an elementary idea which represents the 'whole.' As this whole is divided into parts by adding information, the parts are numbered (or named) after the whole. For instance, the first-level child nodes of the root nodes in a tree will be called first, second, third, etc. but within the context of the first node. The children of the first-order child nodes will also be called first, second, third, etc. but only within the context of the first-order child nodes. This approach to numbering objects leads to a hierarchical object naming.

A fourth-level child object will now be named A.B.C.D, where A represents the first object's number (and should always be 1); B represents the order of the object within the divided set created from the first object, and so forth. Hierarchical numbering conventions are common in everyday space and time notions. Such numbering is also used for identifying nodes in the computer network called the Internet. Of course, in the Internet, the numbers assigned to devices do not signify their meanings or properties. For instance, a search engine does not have an Internet address that signifies the fact that it is a search engine; neither does a social networking Internet address signify the fact that it is a social network. The mapping between Internet addresses and network nodes is established using a Domain Name Service (DNS) which maps names such as www.google.com or www.facebook.com to an Internet address. A computer will resolve these names into Internet addresses before it connects to these nodes. While we understand the significance of these names, the computers do not. In that sense, the names are treated semantically by us, but not by the computers themselves.

In the semantic scheme, the addresses of nodes would directly be representative of their meanings. Some innovative approaches to redefining the internet such as Named Data Networking (NDN)[26] propose to take precisely this approach by naming nodes by their properties and meanings rather than with arbitrary numbers. The deeper problem, however, is that even if we did name these nodes using meaningful names rather than Internet addresses, the machines would still not understand these names because the underlying technology in the machine still works based on physical properties rather than meanings. These machines (computers) don't treat the location of an object in space as indicative of its meaning. For instance, a computer will not store the value of a variable called 'temperature' at a location called 'temperature,' because computer memory is defined physically and not semantically. For computers to understand meanings, the computer would have to employ a semantic technology in which computer memory locations are named (or numbered) by meanings. In such a technology, memory locations would have to be defined hierarchically (current computers define addresses in the memory using a flat addressing scheme).

Such kinds of computers can be built using quantum objects when quantum theory is interpreted semantically and objects locations (numbers) are treated as indicative of their meanings. In such a computer, it is not necessary to translate www.facebook.com into its Internet address, and the computer will know that the name of the address represents a social networking service. When such computers manipulate numbers, they will be computing meanings (all computing is eventually a manipulation of numbers, although numbers currently represent quantities and not meanings). Such a computer can be said to be *thinking*, although it will not be *conscious*. The computer will also not *perceive* meanings or have an *understanding* of what it is doing, although what is doing will correspond to the brain processes during the act of thinking and it will be a material representation of the thinking processes in humans. Such computers will also not be able to create programs or new information on their own. However, given an abstract input (like 'social networking') they will be able to produce the desired service.

Obviously, such technology depends on a prior development of a physical theory of meanings, which in turn depends on the development of a mathematical formalism which describes space and time as a hierarchy of locations such that the locations denote meanings. In a hierarchical space, two objects physically close may not be semantically close, just as we can keep two computers physically close to each other although one serves games while the other serves music. Physical proximity in space will not be a semantic proximity. Rather, all music playing computers will be semantically closer than the machines which serve games. The path of causality between such objects would, accordingly, not be the straight line as the shortest distance between points in a flat space. Rather, the path of causality will be the path that traverses the semantic hierarchical tree of nodes. The time taken in this causal interaction would depend on the semantic rather than the physical distance. Thus, two semantically similar objects can communicate faster because their semantic distance is small (as they require fewer hops on the tree). Similarly, semantically different objects will require more time to communicate or interact causally. A commonplace example of this fact is illustrated by the fact that two people require considerably more time to communicate if they have radically different ideological backgrounds, even though they may be physically close. Conversely, people with similar ideological backgrounds can communicate quickly even if they are physically quite a distance apart.

Communication is an act of object transformation, which requires a notion of causality different from current science. In present science, causes modify the physical states of objects, not their meanings. The physical causality is supposed to modify the object like a ball hits another ball, and the balls travel in a straight line (unless modified by other causes). In semantic causality, the causes will modify the meanings (and thereby the physical states) but these causes will not travel in a straight line, because the straight line path is not the shortest path between two objects. The causes will rather travel up and down the semantic hierarchical tree, which is the shortest path between two symbols that represent meanings.

Such a physical theory obviously requires a new view of space and time. As the numbering scheme A.B.C.D shows, B is a subdivision of the entire space of possibilities represented by A; C is a subdivision of the space of possibilities denoted by B; and so forth. A, B, C, and D are therefore not infinitesimal *points*. They are rather closed regions of space, which can be increasingly subdivided into parts. Similarly, when A.B.C.D denotes a time instance, A, B, C, and D are not infinitesimal instants. They are rather closed durations of time, which can be increasingly subdivided into smaller parts. A closed region of space represents a spatial boundary within which locations are given semantic names, quite like the locations within a house are given names such as kitchen, bedroom, bathroom, study, etc. Similarly, a closed region of time represents a temporal boundary within which durations are given semantic names, quite like durations in a day are called morning, afternoon, evening, and night.

In a hierarchical space-time, space extensions are closed and time durations are cyclic. All locations and instants have a distinct meaning and are called by names consistent with those meanings. That is, we cannot change the names of these locations and instants without changing their meanings. It follows that a hierarchical space-time will forbid coordinate transforms, since transforms in this space-time have the potential of mapping a symbol not-P to a coordinate value P creating a contradiction. Even if the transform doesn't create a logical contradiction, it can create a semantic discord, such as by calling a yellow object by the name 'purple.' A hierarchical and semantic space-time therefore is an absolute space-time. Of course, we could still perform coordinate transforms, but the knowledge produced by these transforms will be inconsistent with reality, and that inconsistency can be empirically verified.

Precedents of such hierarchical space-time models currently exist and are well-known as the mathematical sets called *fractals* which represent repeating patterns at every scale[27]. However, a hierarchical space-time is not necessarily a repeating pattern.

Carbon Dating in Cyclic Time

One obvious criticism of the above described cyclic view of evolution can be that if the evolution was indeed cyclic then we would have found fossil evidence of the same types of animals dating back to different periods in which they existed. Such fossil remains don't seem to exist today. How can then we claim that species are indeed created and destroyed and then recreated? While this argument is generally valid in the context of current fossil dating techniques, some discussion is in order to understand how these techniques would change if the fundamental theory of matter—quantum theory—based on which the dating techniques have been developed is itself modified to incorporate the notion of cyclic time. But, even before we look at that shift, it is important to examine some basic assumptions about how fossils are interpreted as age estimates.

In the following discussion, I will use the carbon dating technique as an example but this discussion also applies to other radiometric dating techniques. Carbon dating measures the ratio of two carbon isotopes—the stable ^{12}C and the radioactive ^{14}C. The radioactive ^{14}C isotope is generally supposed to be created in the upper atmosphere of the Earth when cosmic rays (atomic nuclei traveling at very high speeds) bombard the stable ^{12}C. The stable ^{12}C gains two additional neutrons and becomes ^{14}C. This ^{14}C now reacts with oxygen to form $^{14}CO_2$ which comes down to Earth through rain, is absorbed by the plants, and then consumed by herbivores. The ratio of ^{14}C to ^{12}C is generally observed as roughly being 1 to 1 trillion. Since the ^{14}C is radioactive, it decays over time and if the plant or animal is not acquiring any further radioactive $^{14}CO_2$ then the percentage of ^{14}C in the plant or animal body would continuously reduce. Since ^{14}C has a certain half-life to decay into ^{12}C, the dating method uses the percentage of observed ^{14}C to measure the age.

There are a few key assumptions involved in this dating technique. First, the method assumes that the rate of ^{14}C isotope creation in the upper atmosphere remains constant over time, because if the rate would change over time then age estimation would differ. Second, it must also be assumed that different parts of the Earth's atmosphere are producing and removing the ^{14}C isotope at the same rate, because if the rate differed then the estimated age in different parts of the world would also differ. Third, the method assumes that the half-life of an isotope is constant, because if it wasn't constant in time then the percentage of the ^{14}C isotope relative to ^{12}C could not be useful to determine the age. Each of these assumptions is drawn out of the idea that the inorganic nature is in a steady state and that time passes linearly. If the total rates of production or drain of the ^{14}C isotope varied with time or space, then it would be very difficult to use radioactive decay to determine the age of animal fossils. Similarly, if the rate of radioactive decay changes with time, then it would be difficult to use radioactive data to determine a fossil's age.

The key question therefore is: Does the rate of radioactive decay and that of cosmic rays producing ^{14}C change with time? Does the production of the ^{14}C isotope and the radioactive decay rate change with location on Earth? Today there is evidence for the periodic changes to the rates of radioactive decay[28]. There is also evidence for 1.5 to 2-fold century-timescale changes in cosmic rays[29]. The effects of cosmic ray activity on weather, cloud formation, and a host of other natural phenomena on earth are also speculated[30].

Rupert Sheldrake has similarly argued[31] that many things that we currently consider "constants" of nature—such as the speed of light—are in fact observationally known to vary over time. Sheldrake concludes that the laws of nature are simply "habits" of nature rather than something universal and unchangingly eternal.

Often the key difficulties in interpreting such experimental data is the lack of a theory which can help explain the observations. Are these observations statistical deviations in experiments or do they represent new phenomena that current theories cannot explain? The evolution of science is often catalyzed by a discord between theory and observation. The established theories continue to treat the observations as circumstantial evidence until a new theory comes around and explains them better. Therefore, the interpretation of data—i.e. whether the data requires a new theory—cannot be settled without the actual formation of new theories. It is only in the presence of a theory that scientific observations can be correctly interpreted and it is only in the presence of scientific observations that the complete truth of a theory can be accurately established.

The idea that space and time are hierarchical and closed represents such a thesis. This idea is primarily drawn from the foundational problems in physics, mathematics, and computing although it has profound applications to biology, as we have discussed. One key application of this idea lies in the problem of using radiometric dating techniques to determine the age of fossils, and thereby the time at which the animals—whose fossils are being examined—lived. If space and time are not homogeneous and isotropic (as current physical theories postulate) then the basic assumptions about the universe being in a steady state would be false. We would instead have to assume that the universe is in a state of oscillation. There is evidence available today about these oscillations, so the key scientific need is the formulation of a theory that explains the changes.

The oscillation problem makes the fossil dating method indeterministic. For instance, if you see a burning candle that is only 2 inches long, should you assume that it was originally a very long candle that has been burning for a long time? Or should you assume that the candle was indeed short to begin with and has been burning only for a short while? Both conclusions are consistent with the observations, and knowing the current candle length is not sufficient to know the original length. Likewise, by looking at the current length of the candle we cannot know how long the candle has been burning for, if the rate of burn in the candle changes with time, or if the length of the candle depends on the time when it was created.

Similarly, knowing the current strength of ^{14}C isotope cannot help us determine the age of a fossil if indeed the rates of production and decay of the isotope change with time. A completely different method of estimating ages of fossils now needs to be developed.

This new method, I believe, can emerge from a semantic view of quantum theory. Recall our earlier discussion on fossils which implicitly treat rocks as information about the past, like a book that describes another reality. The book and the reality are connected not physically, but through the ability to *name* objects; for instance, a book can describe Mount Everest, and while the mountain and the book are different things, the book refers to the mountain through a *name*. There are two essential features needed in matter to describe other material objects: (1) there is an absolute naming convention by which material objects can be referred to, which in turn implies a unique coordinate reference frame, and (2) these names can be encoded in matter as some physical properties. My earlier work on quantum theory[32] describes such an interpretation of quantum theory in which objects can refer to other objects, because these objects represent names and concepts by which things can be described.

In a semantic view of nature, atomic objects can refer to the other objects by calling those objects through a hierarchically defined name within an absolute coordinate system. The referenced object may not currently exist, but it will be called by a unique name by which the space-time event in which that object existed is called. The quantum object will therefore encode the object's name even though the object does not exist. By interpreting the quantum physical state as a name of the referenced object, and then converting that name into a space-time event based on a universal coordinate frame that a semantic quantum theory will define, the quantum state in the fossil can indicate the exact place and time (space-time event) at which the fossil was created from some animal remains.

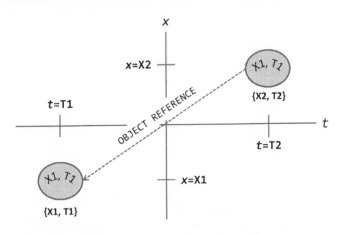

Figure-19 Referencing the Past in the Present

This method of determining the time and place at which the species existed in the past is far more accurate than the methods that rely on the decay of ^{14}C or other radioactive isotopes, because it does not involve assumptions about constant rates of isotope creation and depletion resulting in a steady state atmosphere. Rather, the new approach attributes the same causal mechanisms of cyclic evolution to the atmosphere as it does to species evolution.

In summary, the empirical confirmation of the cyclic model of evolution itself requires reinterpreting the fossil data based on a different theoretical foundation for interpreting quantum experiments, and for explaining the observed oscillation in the atmospheric conditions which determine the radioactive isotope density.

5

Comparative Analysis

His theory had, in essence, preceded his knowledge—that is, he had hit upon a novel and evocative theory of evolution with limited knowledge at hand to satisfy either himself or others that the theory was true. He could neither accept it himself nor prove it to others. He simply did not know enough concerning the several natural history fields upon which his theory would have to be based.

—*Barry Gale*

Intelligent Design (ID)

ID proponents argue that natural processes (random mutations and natural selection) cannot account for the production of biological organisms because these have a degree of complexity which cannot be produced through incremental steps. ID proponents argue that this complexity can only be produced by an act of design by some intelligent designer. It is generally supposed that this designer is God, since other forms of intelligent life require biological complexity which only God could have created. The crux of the argument in ID therefore rests on the notion of irreducible complexity (IC). ID advocate Michael Behe, for instance, argues as follows[33].

> *By irreducibly complex I mean a single system composed of several well-matched, interacting parts that contribute to the basic function, wherein the removal of any one of the parts causes the system to effectively cease functioning. An irreducibly complex*

system cannot be produced directly (that is, by continuously improving the initial function, which continues to work by the same mechanism) by slight, successive modifications of a precursor system, because any precursor to an irreducibly complex system that is missing a part is by definition nonfunctional. An irreducibly complex biological system, if there is such a thing, would be a powerful challenge to Darwinian evolution.

The evolutionist argument against IC is that a system from which some parts were removed may not function exactly as the previous system, but it would function in some way. For instance, if we remove the air conditioner in a car, the car will still work although it will not cool the passengers. We might then partially or entirely remove the seats and the car will still be able to transport passengers, although perhaps with great inconvenience to them. We could then remove the brakes and the car will still move but endanger the driver on a busy road, unless the driver knows how to maneuver the car just with the steering. We can then even remove the steering and the car will still move although this will certainly lead to an accident. These examples illustrate that the function of a car grows incrementally and there is nothing irreducible about it.

However, there are also many examples in which each of the above mentioned functions would completely stop with a minor change. For instance, if we remove the battery from a car, then the car will not start. If we lose the keys to the ignition, or the wire to the ignition is cut, then the car won't start. If the pipe transporting the fuel to the engine is broken, the car won't start. If the axle connecting the wheels is jammed, then, again, the car will fail to start.

So it seems that there are many parts of a car (e.g., steering, brakes, seats, and air-conditioning) which enhance the function of a car but are not critical to its basic functioning. There are also equally many parts of a car (e.g., battery, ignition, fuel pipe, and axle) which are absolutely critical to the car behaving as a transport vehicle. The irreducible complexity in the car, as far as its essential working is concerned, lies in the critical functions working together as expected. This critical assembly may be called the IC in the car.

However, an evolutionist can now argue that each of the critical functioning parts can itself be conceived as comprising absolutely critical and non-critical parts. For instance, the car battery can be broken down into critical components such as lead and acid, and non-critical components such as a non-corrosive sheath. The wire that connects the ignition to the engine can be broken down into a critical component such as a metallic electrical conductor and a non-critical component such as a non-conducting polymer covering. And one might argue that this division of functional components into critical and non-critical parts can be carried out indefinitely. The acid and lead in the battery, for instance, might contain impurities that reduce their electrical voltage and the metal conductor might have impurities that reduce its conductance. They will perform their functions, although the functioning can certainly be improved.

The evolutionist will now conclude that material configurations can evolve through random mutations and natural selections, making incremental improvements upon imperfect combinations.

Where is the problem in this line of argumentation? The problem is not in that the working of the whole cannot be reduced to the working of the parts, but in the *arrangement* of the parts to form a working whole. For instance, a battery connected to the car steering or the car wheels will not drive the car, nor will the ignition connected to the axle produce a working car. Given a set of N parts, there are N^2 ways in which these parts can be interconnected. Each of these N parts, in turn comprises some M parts, which are in turn built from another P parts. The parts too can be interconnected in M^2 and P^2 ways. As we saw earlier, the total number of possible combinations of the parts (if the parts are *a priori* real) is exponentially larger relative to the total number of parts. The various possible combinations of the parts represent different matter distributions and the act of dividing these parts into a structure is the act of drawing boundaries in space-time. As we saw earlier, there is no physical theory that predicts the actual matter distribution in nature; all physical theories only predict all possible distributions.

We also saw earlier how the idea of random mutation is only allowed by current physical theories because these theories are causally incomplete. Matter can be distributed in an ensemble in many ways and a physical theory does not pick a particular distribution. The act of distributing matter represents a choice of how the universe of individual particles is divided into parts through boundaries. Given N objects, there are 2^N possible boundaries, and each such possible boundary and matter distribution can be thought of as a random mutation. However, natural selection cannot pick out amongst these mutations because natural selection only looks for 'compatibility' between an organism and its environment and each of these 2^N possible mutations is already mutually compatible. Natural selections themselves require the presence of boundaries in nature along which compatibility can be tested and ensured, and there are 2^N different ways in which boundaries can be drawn. Thus, while natural selection will eliminate many incompatible combinations between organism and environment, there are still immensely large numbers of combinations that natural selection will allow.

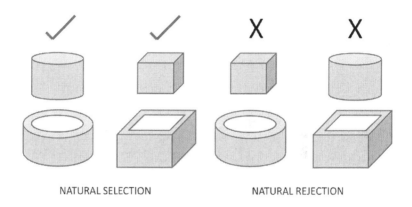

NATURAL SELECTION   NATURAL REJECTION

Figure-20 Natural Selection as Peg-Hole Compatibility

One simple way to think of natural selection is the matching between pegs and holes. Natural selection will eliminate a square peg in a round hole and a round peg in a square hole. However, natural selection will allow a square peg in a square hole and a round peg in a round hole. The indeterminism of physical theories is that there are infinitely many possible distributions of permissible peg and hole combinations. For instance, we can have an octagonal peg in an octagonal hole; or a hexagonal peg in a hexagonal hole. If we have picked up a particular type of peg or hole (by drawing the boundaries in nature) then the corresponding hole or peg can be naturally selected. However, there is no theory that defines how the first peg or hole would be naturally created. The mechanism for subsequent peg and hole creation is identical to the first peg or hole creation. But there is no physical theory that tells us how the first peg or hole would be created automatically because all physical theories are silent on the selection of specific matter distributions in nature.

Evolutionary theory assumes that there are some predefined holes into which the pegs must fit. The modification of the peg to fit the hole is random mutation and the compatibility between peg and hole is natural selection. However, there are several problems in thinking in this way. First, how did the holes take a form? Second, in the search for compatibility which amongst the hole or peg should be modified? Third, there are infinitely many ways of arriving at such compatibilities which are consistent with physical theories; there is no physical theory that picks out a peg-hole combination.

ID claims that each peg or hole has some IC which makes it a particular type of peg or hole (e.g., square or round). This idea is questionable. While the uniqueness of the peg or hole shape cannot be denied, there is nothing that prevents us from converting a square peg into a round peg or vice versa through incremental steps. The problem actually is that there are infinitely many peg-hole combinations and the specific combination is not determined by any physical theory. Only if we assume a specific hole shape can we know the peg shape that will fit into that hole, and natural selection assumes such holes and tries to mutate the pegs to fit them. The choice of the original hole, however, itself begs the question: How was the original hole created, which in turn shaped the pegs?

The correct way of think about evolution is not to assume that there are predefined holes which then modify the pegs until they fit. The correct way to think of evolution is that it creates mutually compatible pegs and holes simultaneously. That is, the peg and the hole are either defined at once, or not at all. This definition of pegs and holes is underdetermined by physical theories and therefore a new mechanism is needed to explain them. The new mechanism can be *information* which distributes matter in mutually compatible collections (the simplest collection in this case would be a pair). This information is built from distinctions comprising opposites which allow the environment and the organism to be locked in mutually complementary relations. However, these relations are natural entities and can be studied in science and described mathematically. They are not physical objects or observable entities, however.

We can conceive boundaries in nature, but we cannot perceive them. Boundaries in nature can be mathematically described and their effects can be empirically measured. But we cannot observe the boundaries themselves. In that sense, the information that creates mutually compatible pairs is natural but it is not physical. We have to suppose that there is another kind of material reality—which we call concepts—that distributes matter in space-time. These concepts are understood by the mind and they can be created by the mind. The manner in which these concepts interact with the world is however not understood in current science. That, I tend to suppose, represents an area of future development in science.

ID attributes design in nature to some intelligent designer. A common interpretation of this intelligence is God, and this idea is problematic because it precludes the possibility that the organism's mind may itself create new design. It becomes further problematic if the design undergoes change according to natural laws, because then we would have to suppose that God is governed by natural laws. Of course, this isn't necessarily a scientific problem; whether God is governed by natural laws is more a theological question than a scientific one. The scientific question is whether we can justify the idea of ID based on IC. I believe that establishing the connection between ID and IC needs a further development of physical theories.

We saw earlier how irreducibility implies irreversibility and non-linearity which is observed in the statistical mechanical formulation of thermodynamic phenomena. To the extent that biological beings are governed by irreversible and non-linear theories, the irreducibility is undeniable. However, this irreducibility cannot be easily mapped to molecular mechanisms because the current theory describing these mechanisms is a linear and reversible theory. So the notions of irreducibility and linearity are mutually incompatible. As we saw earlier, this paradox can be resolved through a theory in which non-linear phenomena are byproducts of less-than-maximal information while linear phenomena involve maximal information. However, this shift itself requires an informational view of nature in which space-time distinctions are produced by encoding more and more information, and any object that still has the potential to encode more information is a non-linear and irreducible system.

The relation between ID and IC is therefore very tenuous currently, especially if this connection is based on molecular mechanisms. This relation can however be justified in a formal theory that treats matter distributions as an outcome of information creation.

ID theories build upon the idea of objects in physics which exist before being organized into structures. For instance, if applied to the construction of household furniture, ID would imply that the planks of wood exist *a priori* but the process by which these planks are organized into a chair, table, or bed needs an intelligent designer. The problem now is that the plank of wood itself has structure which could be broken down into smaller parts, and each part further divided into even smaller parts, until we arrive at the smallest sub-atomic particles. Should we postulate design for tables and chairs and not for the atoms and molecules? If yes, why would we permit that discrepancy? It is not clear when the structure in ID emerges. For instance, are atoms also structured by design or does this structure emerge solely at the level of molecular mechanisms?

The semantic view differs from ID in the following fundamental sense. In the semantic view, the planks of wood that make up the chair and the table are different. This difference is not just due to the designer nailing together some planks to form a chair or table. Rather, the difference is also at the level of atoms, which cannot be defined except in relation to the entire whole. Thus, we cannot reduce the 'table leg' to a plank of wood that exists identically outside the table. Rather, the atoms in the table themselves are defined only in relation to the table. To describe these atoms, we need to use a hierarchical notion of space-time in which the idea of table is divided into the ideas of 'leg,' 'top,' 'drawer,' etc., which are further divided into even smaller parts, to greater and greater levels of detail. Now an atom is not something that can be defined outside the table. Rather the atom must be defined as 'part-of-a-table' such that it is essential to speak of the table before we can speak of its parts.

ID attributes design to the more complex parts of the biological system, although not to the simpler parts. This is the main flaw in ID because it is impossible to identify the level of complexity at which the design suddenly becomes relevant. Furthermore, by treating the complexity as an act of design, ID dissolves the need to explain and predict this complexity, which are basic goals in science. In effect, ID describes the complexity in biology as a limit to science: the complexity can be described but cannot be explained naturally.

The semantic view, on the other hand, does not treat complex and less-complex objects differently. It views complexity as a general problem of trying to explain the relation between part and whole, which is not restricted to biology but spans across mathematics, computing, physics, and other areas of science. This treatment entails a very general theory of complexity that depends upon fundamental changes to how numbers, objects, and programs are described, before biological systems can thus be described. Finally, only such a treatment, I believe, can provide a predictive theory of evolution; neither current evolutionary theory nor ID are predictive theories. In so far as the goal of science is explanation *and* prediction, both evolution and ID are incomplete scientific theories: they provide an explanation of the biological complexity but not predictions about how this complexity evolves through natural laws.

Punctuated Equilibrium

One way in which PE as a model of biological evolution—originally proposed by paleontologists Niles Eldredge and Stephen Jay Gould in 1972—can be understood is to compare it with idea of "paradigm shifts" (PS) proposed by Thomas Kuhn as the manner in which scientific ideological change comes about. Both Kuhn and Gould held that change comes about not via incremental variations (in biology or in ideologies) but through dramatic and large-scale sudden shifts.

Scientific development depends in part on a process of non-incremental or revolutionary change. Some revolutions are large, like those associated with the names of Copernicus, Newton, or Darwin, but most are much smaller, like the discovery of oxygen or the planet Uranus. The usual prelude to changes of this sort is, I believed, the awareness of anomaly, of an occurrence or set of occurrences that does not fit existing ways of ordering phenomena. The changes that result therefore require 'putting on a different kind of thinking-cap,' one that renders the anomalous lawlike but that, in the process, also transforms the order exhibited by some other phenomena, previously unproblematic[34].

My own field of paleontology has strongly challenged the Darwinian premise that life's major transformations can be explained by adding up, through the immensity of geological time, the successive tiny changes produced generation after generation by natural selection[35].

This comparison between the evolution of ideas and the evolution of living beings might seem facetious to some, but this will indeed be the conclusion that one will be led to when the current problems of indeterminism, incompleteness, incomputability, and irreversibility in physics, mathematics, and computing are solved. As I have detailed earlier, the only consistent and complete way to solve the problems in modern science is to think of material objects as symbols of meaning, which are produced from meaning. The structure and order in matter encodes meanings that can exist even outside matter. While we cannot directly describe those meanings in science, we can certainly describe their material embodiments using a new kind of physics, mathematics, and computing theory that views nature as comprising hierarchical space-time events.

The difficulty in both Kuhn's and Gould's theories is however that they did not provide the *mechanism* or *explanation* of this change. This becomes particularly relevant when both kinds of evolution obviously seem to involve both incremental and dramatic changes. Which kind of theory would explain both kinds of changes? When is evolution incremental and when is it dramatic? The conflict between PE and other theories of evolution rests on this foundational inconsistency in explaining two different types of change.

This problem is easily demystified if evolution in nature is seen as the nested hierarchy of various time cycles. In physical theories, complex evolution patterns are routinely reduced to their component frequencies using a technique called a Fourier transform. Figure 21 illustrates an example. The left side of the figure shows the individual wave patterns being combined and the right side show the result of that combination. When the waves interfere constructively, they cause large-scale changes and when they interfere destructively they result in incremental and slow-paced changes.

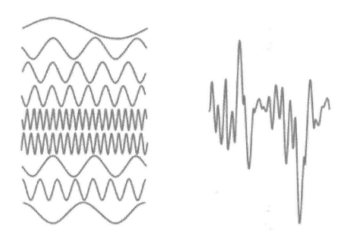

Figure-21 Incremental and Sudden Changes

Of course, if these waves were universally extended, then the same patterns of evolution would be observed at different times at different places. This isn't obviously the case with observed facts. The interference of oscillations therefore cannot be described in terms of universally extended oscillations. They must rather be described in terms of different interference patters in different locations. This is possible if space itself were divided into smaller regions such that each location has a unique interference pattern arising out of a unique set of oscillations associated with each location. This observation asserts the need to think of space as being divided into hierarchically organized regions, such that each region has an oscillation pattern and the regions nested within the larger region incur the interference between that region's oscillation and the oscillations of all the regions that are larger and therefore nest it.

This method of describing the evolution of nature is consistent with quantum theory which also describes material phenomena in terms of waves, although the quantum theoretic description needs to be extended to macroscopic objects for us to think in this way.

Current quantum theory is selectively applied to atomic reality under the view that the macroscopic world is classically deterministic and should therefore be described using Newton's laws. The selective application of quantum theory to the atomic world is also necessitated by the problems of probability in quantum theory, because current quantum theory describes nature in terms of probability waves which must be 'collapsed' to create a definite reality. We don't suppose that the macroscopic world is probabilistic; the macroscopic objects—living beings included—don't appear and disappear, and we do not suppose that our hands and legs only have a certain probability of being in their place when we wake up next morning. Thus, while the above type of description of evolution of macroscopic systems is consistent with the foundational ideas in quantum theory, the application of the theory to macroscopic phenomena in turn begs a solution to the problem of probabilities.

If the macroscopic universe were treated according to quantum theory—which would be possible if the problem of probabilities was solved using an informational view of nature—then it would be possible to think of systems as waves. The periodicity of waves would represent a cyclic evolution of the macroscopic system. Further, if these waves are defined hierarchically—because the system is nested inside larger systems—then the superposition of the waves could be used to explain rapid and slow change through the same theory. PE and slow evolution would then be a consequence of the same type of theory, rather than two different theories.

The conflict between PE and gradual evolution is therefore a consequence of describing matter as independent objects rather than as systems, and then describing these objects in terms of classical concepts rather than quantum theory. The formation of new species involves dramatic changes to the ecosystem as PE already illustrates (and as we previously discussed—the ecosystem has to change as a whole to create new species and individual changes without a concerted ecosystem change would always be reversed). Therefore PE is the correct explanation of the formation of new species than slow evolution. Even slow evolution requires the ecosystem evolution, rather than the evolution of an individual species. Both these types of evolutions, therefore, require the biological ecosystem to be treated as a system that evolves in concert. PE and slow evolution appear to be contradictory in the current dynamical view of nature, but they will, I believe, be aspects of a single theory of systemic evolution within a new kind of space-time theory.

Morphic Resonance

Rupert Sheldrake advocates a new vision of science in which fundamental *forms* underlie physical phenomena. These forms, Sheldrake suggests, persist even when the objects that embody them are destroyed. Since the forms are not destroyed when the objects are, the forms appear again and again in different objects. Sheldrake also claims that once these forms have been created, it becomes easy to create them at different places in the universe; for instance, if a certain type of crystal was once created in one part of the world, it would then be easier to create the same crystal in other parts of the world. This ability to transfer the forms from one location to another Sheldrake attributes to what he calls *morphic resonance*. The basic idea is that forms exist as fields of vibrations and these fields can resonate with other similar fields in other locations, thereby making it easy to transfer the forms from one place to another.

There are several similarities between the idea of forms as described in morphic resonance and the notion of semantic matter described earlier in this work. As I have noted earlier, material objects are produced from ideas or meanings and these ideas can be said to exist as forms. These forms can be observed in space-time by the senses when the senses are more abstract than the forms. However, when the forms are more abstract than the senses, then these forms cannot be perceived by the senses and newer kinds of perception mechanisms must be postulated to understand these forms. Since material objects are produced from ideas, there is some similarity between the notion of morphs and the semantic view.

The difference between morphic resonance and the semantic view lies in how ideas are replicated or recreated in different parts of the world. While Sheldrake attributes this replication to a 'resonance' between these forms, in the semantic view, the replication is an effect of time. In the semantic view, ideas exist as possibilities which are actualized as realities in due time. Which ideas are realized when depends on the *type* of time (time and space are types in the semantic view). It is therefore correct to suppose that if a crystal was once created at one particular place then it would be easier to observe its production at other places as well. However, that is not because the idea itself is transferred from one location to another, nor is it due to a resonance between the forms at two locations. Rather, the new idea is independently produced at the other location because time is typed and causes the production of the same forms or things at different locations. To recall Victor Hugo's famous statement, nothing is as powerful as an idea whose time has come.

In both morphic resonance and the semantic view, multiple instances of a given type of species can be produced in quick succession even though these instances are not descendants of a single organism. In the theory of random mutation, even the production of a single organism is a very long drawn and improbable event, so the simultaneous advent of multiple instances of such organisms which are not direct descendants is even more unlikely. Both morphic resonance and the semantic view allow the simultaneous creation of many instances of the same species because all these instances are essentially instantiations of the same idea which previously existed as a possibility but could then be expressed as a material arrangement. The crucial difference between morphic resonance and the semantic view is that the mechanisms underlying the simultaneous production of instances of the same species are quite different. In morphic resonance, a form is first produced randomly and then quickly replicated due to resonance. In the semantic view, the forms are independently produced because the production of the form is caused by the cyclic and typed nature of time. Essentially, when the time has changed its quality, many copies or instances of the same idea would be produced either simultaneously or in quick succession. For these ideas to survive in an ecosystem, they generally aggregate, and morphic resonance can assist in their aggregation.

The ideological evolution in society follows a similar pattern. The evolution begins when several individuals simultaneously become dissatisfied with a certain way of thinking. They seek alternatives and a particular type of idea gains stronghold in many different minds. Similar minds then come together and communicate these ideas, creating more synergy and harmony between the ideas, which is a similar to the notion of morphic resonance, except that this resonance involves acts of communication to spread the ideas. If morphic resonance alone was sufficient to explain the spread of ideas, then there would be no need for individuals to communicate.

Sheldrake justifies morphic resonance based on the sense that many people often have of being stared at, or the fact that some dogs become aware of their masters coming home, well before they have actually arrived home. Sheldrake attributes this awareness to a morphic field that exists in the two individuals and resonates to create the awareness that someone is staring at us or the awareness of masters coming home in dogs. This idea is however quite different from the other ideas about crystals automatically replicating or the simultaneous production of the same species members. The difference is that in both the above cases, there appears to be some sort of telepathic communication between different living beings although we can't suppose such a communication in the case of crystals or the simultaneous instantiations of the same species.

Sheldrake's thesis raises a question about the mechanism underlying telepathic communication: Is this communication due to morphic resonance? Or is there another scheme involved in it?

I believe that the problem of telepathy (as a way of communicating ideas) cannot be resolved without first defining the mind. All communication is currently defined as a transfer of energy. Resonance is a concept that originated in classical physics where a system has a natural proclivity to vibrate at a certain frequency. However, there has to be an energy transfer to the second system to put it into that state. Similarly, in quantum theory, resonance techniques (such as Electronic Spin Resonance or ESR and Nuclear Magnetic Resonance or NMR) are observed when a force field (e.g., an electromagnetic field) is applied to a quantum system. This means that the force field dissipates its energy *before* another system resonates. The temporal ordering of dissipation and absorption is currently described as all force fields being *local*—the energy is said to be emitted as a photon and then absorbed at another particle after traveling in space. Classically, we think that whichever particle the photon hits first would automatically absorb this photon. But, quantum mechanically, only those particles whose next higher state is identical to the energy of the photon would actually absorb it.

Potentially, however, if two individuals are very far part in physical space, there are many physical objects between them which can potentially absorb the emitted energy. Therefore, even if we suppose that some energy is emitted from one object to another, it would be absorbed by the intervening particles. This makes the problem of telepathy very intriguing and very difficult to solve.

The semantic view I described earlier brings two important differences to the current way of thinking about resonance. First, communication is not a transfer of energy but a transfer of meaning; this transfer causes some physical changes which we currently *model* as a transfer of energy. For two systems to communicate effectively, they must be similar in some sense to begin with. This is commonly understood as the fact that if two people have to start conversing effectively, they must begin from a common ground that they both understand. When quantum objects are described semantically, the similarity between these objects would also be described in terms of their meaning rather than their energy state. That in turn would mean that if a quantum can emit energy E_x, and there are many quanta that can potentially absorb E_x, there will only be certain quanta that will actually absorb it. The quanta that absorb this E_x would represent a similar meaning besides just the physical ability to absorb the energy. Note that a quantum in different contexts can represent a different meaning and the meaning of a quantum is not identical in all ensembles even though their energy states are the same. Thus, for a quantum of energy to be emitted and absorbed, there must be a semantic similarity between the source and the destination. When quantum theory is understood semantically, it will predict which sources and destinations exchange energy.

There is a big difference between classical and quantum theory vis-à-vis their descriptions of energy transfer. In classical physics, the energy is simply emitted and it travels outwards from the source; when it 'hits' another object, the energy is absorbed. Quantum theory instead tells us that even if a quantum is emitted, it will not be absorbed by an arbitrary object, even if that object was in the *path* of that quantum. That quantum can only be absorbed by specific objects whose difference between energy levels matches the energy of the quantum. In a sense, the photon passes 'through' those objects whose energy differences don't match and it is absorbed only by those objects where the energy levels match. This difference between classical and quantum physics is rather profound and not often well understood. Essentially, in classical physics, energy is absorbed by the first object that the energy hits it but in quantum theory the energy 'knows' where it is going even before it sets out.

The flaw in current quantum theory is that it does not accurately predict the actual destination at which the quantum will actually be absorbed although the theory does make probabilistic predictions. If there are N destinations at which this quantum can be absorbed, current quantum predicts a probability of the quantum being absorbed at these destinations but not the actual destination. In a semantic quantum theory, this problem could be solved by distinguishing the particles which have identical energy states by their different semantic states (the same energy state will now represent a different meaning state as part of different ensembles). Now it is possible to speak about which source quantum 'resonates' with which destination quantum. In some sense, this provides a material basis for intentionality because now a particular source only communicates with a particular destination. The manner in which a source 'knows' about a destination (before it transfers the energy to it) is fixed by the fact that they have similar semantic states.

In such a theory, it becomes possible to speak about resonance between quantum states even though they are very far apart. These states can be said to be 'entangled' by their meaning and they can exchange energy due to this semantic entanglement although the energy does not 'travel' through the intervening locations between the two objects, lying on the straight-line path that connects them.

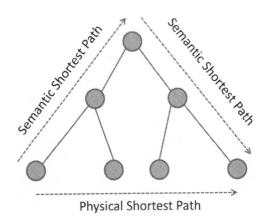

Figure-22 Physical and Semantic Communication Paths

This notion of communication between two objects can further be clarified in a hierarchical space-time where energy is transferred not on a straight line path, but through the hierarchy of locations. These hierarchical locations are *in* the space-time if we look at the space-time physically. But they are also hierarchical if we describe them semantically. This hierarchy arises because locations in space-time are described as types rather than as quantities, and the types are defined in relation to the wholes. The next-hop location is therefore not the immediately next physical location, but the location from which the current location was semantically derived.

Phenomena like telepathy can therefore be understood within science, but it requires us to change the theory of space-time from linear to hierarchical. Accordingly, the notion of morphic resonance that causes the minds to feel connected is possible in a non-physical description of quantum phenomena. While Sheldrake, I believe, has described a new scientific concept, there are several areas in which this concept needs to be further developed outside biology before it can be a reasonable explanation for phenomena inside biology.

The Genetics Conundrum

One of the interesting facts about genetics is that humans have only about 20,000—25,000 genes although far simpler species such as rice has about 38,000 genes. Amoeba carry about 200 times the DNA as that found in humans. If therefore we correlate the complexity of the organism with the number of genes in it, we would conclude that rice grains or amoeba were more complex than human beings. An even more startling fact is seen in the case of different drosophila strains: there is a huge amount of genetic variation across drosophila strains even though a similar type of genetic variation in other cases would have led us to the conclusion that these genes pertain to different species. The human and chimpanzee genomes are about 99% identical. Therefore, if we were to look to the genes to account for the differences between humans and chimpanzees, we would conclude that humans and chimps are 99% identical. These facts about the DNA raise the problem of correlating the genotype and phenotype complexities. If, however, they are not correlated, then how do we explain the gaps between the two types?

An even more troubling fact about DNA has been discovered in recent years. It has been observed that often significant portions of the DNA are actually not being transcribed and translated into proteins. Furthermore, the extent of this non-coding DNA varies amongst different life-forms. For instance, about 98% of the human genome is believed to be non-coding DNA[36] while about 2% of the DNA in certain bacteria is believed to be non-coding. It would appear that if the important part of the DNA was that which actually coded the protein generating information, then non-coding DNA would be dropped by the evolutionary mechanisms as these seem to be unimportant and redundant. Evolutionists have, however, argued to the contrary: they claim that the existence of non-coding DNA indicates the vestiges of evolution left behind by a changing past. It therefore seems that both the existence and the non-existence of non-coding DNA could be justified by an evolutionary mechanism, and the theory is incapable of predicting which of the two is real.

Recent evidence challenges the idea that non-coding DNA is a vestige of evolution and that it has no current role in the organism.

For instance, Peter Andolfatto's research[37] demonstrates that about 40-70% of the mutations arising in non-coding DNA are not passed on to the offspring, suggesting that these genes are actually useful in a sense quite different than the coding DNA. While the coding DNA represents the observable traits of organisms (which evolution claims can be modified to produce organisms more suited to their environment), the non-coding DNA represents that which will resist change. The non-coding DNA is also believed to 'regulate' or 'control' the expression of protein coding genes by allowing repressor, enhancer, promoter, silencer, or insulator molecules to bind to this non-coding DNA thereby affecting the expression of coding DNA.

There is also a growing body of evidence on epigenetics which suggests that the expression of genomic information is controlled by other factors outside DNA. In the canonical example of epigenetics—called methylation—a methyl (CH_3) group is added to the cytosine or adenine DNA nucleotides which effectively deactivates the nucleotide and prevents its expression into a protein molecule. Such methylation enables the same DNA to be used in different cells, even if the cells perform different functions. Effectively, the DNA is like a knife which can be used to cut, pierce, or open in different contexts. The DNA alone is thus not enough; rather, it has to be combined with information about its context to produce a function. This fact has recently become very important in the study of diseases; it has been found that the contextual information changes the manner in which DNA is expressed and hence the functions it performs. And if this contextual information depends on the environment, then the latter plays an equally important role in a phenotype's properties.

Non-coding DNA and epigenetics considerably modify the picture of inheritance as understood in the 1960s. In that picture, all information about the organism resides in the DNA and this information pertains strictly to how the DNA encodes protein information. The organism's complexity is thus correlated to the DNA's complexity; the more complex the DNA, the more complex the organism. Over time, gaps between organism and DNA complexities, the presence of non-coding DNA, and the presence of epigenetic factors have significantly altered that picture. It is no longer possible to think of the DNA as just protein coding information. Rather, we have to look for other kinds of information, including the ways in which this information can be controlled, subdued, enhanced, or modified to produce function, which represents a kind of meta-information. Since both non-coding DNA and epigenetic factors are hereditary, they have an equal amount of role in determining phenotypes.

The problem in genetics at present is how we distinguish between various types of information and understand their different roles. Here, I will use a simple computer (called a Turing Machine) to illustrate differences between types of biological information.

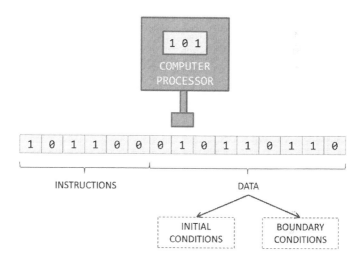

Figure-23 A Simple Turing Machine

Every computer program generally involves a combination of instructions and data. Instructions are what the program does, but these instructions require input data to obtain an output. The input data modifies the behavior of the program, and there are two broad ways in which a program's output can be modified. A program to compute a number's partial factorial, for instance, can be written as follows. The output of this program depends on two inputs—the value of N which is the number whose factorial is being computed and the value of M which is the smallest number in the factorial.

```
PartialFactorial (N, M) = {
    If (N > M) {
        Return (N * PartialFactorial (N-1, M))
    }
    Return 1
}
```

The number N represents an initial condition and the number M represents a boundary condition. If N > M, the program will produce a number N * (N-1) * (N-2) * ... * M. If N = M or N < M, the program will return 1. The number M therefore prescribes the valid values of N under which a partial factorial will be computed.

Every meaningful program has to be finite, and every finite program can be described by prescribing the *set* of numbers that it generates (there are issues of meaning related to how these numbers are interpreted, which I will discuss shortly). For the program to halt in a reasonable amount of time, the values generated by that program must also be finite. The set of possible values that a program can generate therefore represent the *boundary* conditions that the program must follow. To pick a specific value out of the possible values, a specific *initial* condition must be passed to the program. The output of the program is therefore underdetermined unless both initial and boundary conditions have been prescribed.

This fact about computer information can be used to understand biological information as well. Essentially, the part of the DNA that codes the instructions for the production of proteins is underdetermined without the initial and boundary conditions on that DNA code. Non-coding parts of DNA and epigenetic factors can now be understood as representing initial and boundary conditions. Only the combination of instructions, initial, and boundary conditions, represents a complete program. Similarly, only the combination of coding, non-coding, and epigenetic information represents the complete recipe for producing all the biological traits in an organism. Like the output of a program can be modified by changing initial and boundary conditions, similarly, the biological complexity can be modified by the presence of non-coding and epigenetic information.

We can now think of the coding DNA as the initial conditions, the non-coding DNA as the program which transforms initial conditions to outputs, and the epigenetic information as the boundary conditions which constrain this transformation[38]. If the non-coding DNA as the program instructions is modified, the same initial conditions would be converted into different outputs. Some of these outputs could lie outside the constraints placed by the boundary conditions, and either the program instructions or the initial conditions must be modified to meet the boundary conditions. Similarly, if the initial conditions in the coding DNA were modified, the same program could produce outputs that lie beyond the boundary conditions, and the initial conditions or the program must be changed to fit these conditions. Likewise, if the boundary conditions as the epigenetic information were modified, then it would forbid some previously allowed initial conditions and program instructions. These changes would result in the modification of non-coding DNA as the program instructions or to the coding DNA as initial conditions.

It is now easy to see why coding DNA, non-coding DNA, and epigenetic information cannot always be independently changed. There are certain cases in which changes in initial conditions, given some program, will produce outputs within the boundary conditions, and in such cases it is easy to think of mutations to the initial conditions as resulting in the change in output. However, there are also equally many situations in which changes to initial conditions must be accompanied by changes to boundary conditions, if the program conditions are unchanged. And in certain cases, changes to initial conditions must change the program if the boundary conditions remain unchanged. It is impossible to understand and predict which type of information change is related to which others unless all this information is treated semantically. Why some mutations are reversed while others result in new mutations cannot be predicted unless the genome and epigenome are seen as a single semantic system, comprised of different types of information. If we only viewed each molecule as a thing, we would be unable to predict and explain the relationships between the biological information changes.

To note that some molecules represent programs, while others represent initial and boundary conditions, there is a need to assign *types* to molecules. In the physical view of information, this distinction does not *prima facie* exist. Computer architectures for instance distinguish between programs and data (initial plus boundary conditions) by placing them in different memory locations (called text and data sections of the memory). If we treat the entire memory only as a physical expanse of bits, then data and instructions look identical (they are both comprised of binary digits—1s and 0s). The ability to think of molecules as representing different types of information itself requires the ability to apply types to molecules. And if molecules can represent the type distinction between programs, initial and boundary conditions, then their other properties—such as the semantic content of the program, initial, or boundary condition—too must be described as types rather than quantities.

This, as we have seen earlier, needs a symbolic view of nature in which meanings are encoded in matter and objects are viewed as representations of meanings rather than meaningless things. It necessitates a different view of numbers (types instead of quantities) and that of matter, causality, and space-time, which too must be described in terms of types rather than quantities. Types are in turn constructed by dividing elementary memes by those same memes, producing ever more complex ideas and their physical symbols.

The Replication Problem

The primary problem in biology is not that there are many species that need to be explained on the basis of an original species. The primary problem is why there is even one species that is capable of reproducing and passing its traits to its offspring. The question of the diversification of the species arises after there is at least one species that can mutate and pass its traits to offspring. Indeed, the notion of a species is not that there are many individuals of the same type, but that they can replicate and create more of that type (the problem of reproduction). Before biology can address the problem of many species, it should address the problem of a single species, no matter how primordial or primitive that species is. Again, the first problem of biology is not the diversification of the species but the very appearance of a single species. The existence of species may be a fact, but it is on par with any other fact that science is trying to explain—namely that we must find causes responsible for making phenomena happen. Unfortunately, there is no good causal theory today that explains reproduction in a species. Without reproduction there is no species and hence the question of the diversification of species does not even arise. So, before we focus on issues of species diversification, we should focus on a single species.

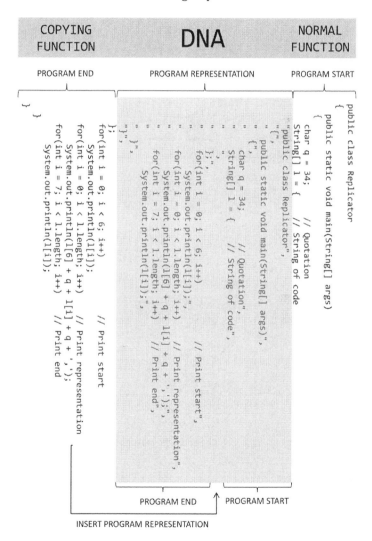

Figure-24 A Self-Replicating Java Program

Reproduction in a species is supposed to occur on the basis of gene replication, which is supposedly caused by chemical reactions. But there is nothing in the theory of chemical reactions that explains how molecules create their replicas. If chemicals are explained by atomic theory, then atomic theory must explain replication. However, there is nothing in today's physics that explains replication.

Figure 24 depicts a simple self-replicating program. This program contains two kinds of information: (a) the program which is executed, and (b) a textual string that *represents* the same program in an 'inactive' form such that it is never executed. While the program instructions are executed by a computer, the string is inactive. The program itself consists of two parts: (a) a normal function and (b) a replicating function. The latter converts the inactive textual representation of the program into a live program. Note that the replicating function would not work unless there is an inactive copy of the combined normal and replicating function to be converted into an active program. And the inactive copy cannot be created unless both normal and replicating functions are already present.

In the case of biological information, this means that the DNA must encode the genes responsible for reproduction although a method of reproduction must exist even prior to the existence of DNA, because otherwise the DNA would never be used. In other words, an active component of the cell must exist which replicates the DNA. But this component must exist before the DNA and the DNA presence is not responsible for the replication. This raises a typical chicken-and-egg problem—which came first: the DNA or the cell capable of replicating, transcribing, and translating the DNA?

We encountered a similar problem earlier in the discussion of the relation between a 'chair' and a 'chair leg.' For something to be a 'chair leg' the 'chair' must exist prior. But the 'chair' cannot exist prior if it is only a combination of 'chair leg,' 'chair seat,' 'chair back,' etc. The need to call something a 'chair leg' leads to a problem of recursion since the 'chair leg' depends on the existence of a 'chair,' but the 'chair' in turn depends on the existence of a 'chair leg.' We also discussed earlier how the only possible resolution to this recursion is if the 'chair' exists prior the existence of its parts. The chair is then divided by adding information to its different parts. The parts of the chair are called 'chair leg,' 'chair seat,' etc. because they are produced from a 'chair'. Of course, the 'chair' in question exists only as an idea prior to being divided into the various parts.

The existence of the DNA further complicates this problem because now a part of the cell encodes the entire cell. In analogy to a human body, which has many parts (quite like the chair), but it also has a part—the brain—which represents the body in the body, the DNA in the cell becomes the 'brain' that represents the entire cell. If we had a body but not the brain, then we could exist like a chair but not communicate our own existence externally. Similarly, if there was a cell that did not have a DNA, it could not replicate itself.

The existence of DNA is like a mathematical set that includes itself, although this inclusion is not that of an exact physical replica of the set. The DNA is rather a *representation* of the idea of the set, which is sometimes called its meaning or *intension*. The set itself includes many things, including a representation of its intension. An everyday analogy will probably help here. Children often play with modeling kits that can be used to build gliders, ships, tanks, or cars. Each such kit has many parts that must be employed in the construction of the glider, ship, tank, or car. But it also has an instruction manual that depicts how to assemble the entire model. The pictures in the manual are information about the model, but they are not the model itself. In that sense, those manuals are part of the total kit, although they are also a representation of the entire kit. The manual is a physical object, but an object that describes the kit. The other parts of the kit can be viewed physically although the manual has to be viewed informationally; it encodes concepts, pictures and names which are things but also descriptions of other things.

It is known from Shannon's Information theory[39] that physical information cannot be compressed unless the information is redundant. However, we habitually compress the information from the world into sentences and pictures which are physically much simpler that the world. How is this achieved when Shannon's theory stipulates that information cannot be compressed except by discarding redundancy? The difference is that in the everyday world we treat information semantically rather than physically. We can never embed the representation of an object inside that object, if both the object and the representation are treated physically. The only way we can embed an object's representation inside that object is if the representation is treated semantically. However, if the representation can be sematic then the object itself can be semantic. The notion of a DNA that encodes biological information therefore itself entails a different understanding of nature—which physical information cannot be compressed without discarding some details, semantic information can be compressed. Now, DNA should not be seen as physical information because in that case the DNA would be incapable of replicating the bodily features. Rather, the DNA must be viewed as semantic information that has been compressed.

The Nature of Biological Information

Biologists routinely use the term 'information' to describe the protein encoding abilities in the DNA. They also, however, routinely reduce this information to the physical states of molecules, which are in turn reduced to the physical states of atoms and sub-atomic particles, which are not described informationally. The information in DNA is supposed to be *about* the processes in the cell, although the sub-atomic particles from which the DNA is constructed are not supposed to have aboutness. The information in the DNA compresses or *abstracts* the information about the world, because if it could not abstract the information then to retain the fidelity of the biological reality the DNA would have to be at least as complex as the world it describes and it could not then be included inside the cell. Abstraction and aboutness are basic properties of semantic information but not of physical information. Information in a physical object cannot be compressed without loss of fidelity, which means that it can never be abstracted into a lesser complex object. Moreover, a physical state can never refer to another physical state.

The use of the term 'information' in biology therefore requires clarifications. What we mean by this information has the semantic connotations of abstraction and indirection, although those ideas cannot be derived from currently physical theories. That doesn't necessarily imply that abstraction and indirection cannot be features of matter, although it does imply that they cannot be understood from current physical *theories*. Semantic information requires physical objects to denote *names* and *concepts*. A name refers to another object and a concept abstracts that object's meaning. Words in ordinary language refer to other objects by names and they denote concepts by abstracting their properties. This ability does not exist in current physical theories, because objects are described independently of the other objects rather than through a relation to the whole system. The whole system is itself ephemeral in current science because the *boundaries* of a system cannot be observed. The shift to semantic information therefore requires the induction of a new scientific idea—boundary—which can be divided in two ways to represent names and concepts or indirection and abstraction.

Such a shift in science is not limited to biology but must arise from even more fundamental shifts in mathematics, computing and physics because the problems of abstraction and indirection are not unique to biology. The ability to refer and represent exists in mathematics when a proposition describes and names another proposition. It exists in computing when a program is given a name and is intentionally connected to the problem it solves. Reference and intentionality are also seen in material objects such as books or pictures that describe and refer to other real world objects. Scientific theories become inconsistent, incomplete, indeterministic, and incomputable, when meanings are reduced to physical states. The problem of meaning therefore has real counterparts in current physical theories although the connection between these problems and the issue of meaning hasn't thus far been clearly understood. A deeper understanding of the problems of abstraction and indirection in other parts of science is needed before biology can be revised to include semantic information. The shift from physical to semantic information will therefore revolutionize all areas of science.

Descartes separated the mind from the body and the properties of indirection and abstraction were pushed into the mind. When the mind is reduced to the brain, these properties are lost. The need in science is to view matter as information about matter, reintroduces the need for the properties of indirection and abstraction into matter. Such a theory of matter will require a hierarchical and closed view of space-time, as we have previously discussed. Nature, in this view, is not an infinite expanse of space-time occasionally peppered with independent objects. Rather, objects are forms in space-time, space-time is closed and hierarchical, and objects and space-time events are produced by adding information to elementary ideas.

To describe the evolution of the universe in this way, science needs to speak about matter in two distinct ways: (a) as objects that are manifest and (b) as possibilities of ideas that are currently unmanifest but could be manifest as objects. Since all objects are essentially representations of ideas, the evolution of objects is not governed by the forces in matter, but by the manifestation of ideas into things and the disappearance of these things back into ideas. For instance, if an object changes state from A to B, the state A hasn't transformed into state B. Rather, the idea of type A has disappeared and the idea of type B has appeared. In the science that describes nature as state evolution, the past, present, and future don't exist simultaneously. In the science that describes nature as the appearance and disappearance of ideas, the past, present and future exist at once (all ideas exist simultaneously) although they are concretized into things and then sublimated back into ideas.

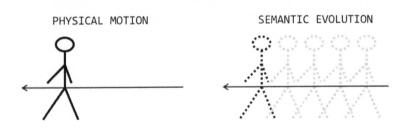

PHYSICAL MOTION   SEMANTIC EVOLUTION

Figure-25 Physical and Semantic Evolution

Think of how classical physics describes a person's walk on the street: a person is a particle whose identity is unchanged throughout the walk, and the particle changes position due to forces or its intrinsic inertia. Now contrast this description of a walk with the image of a walking person which is simulated by blinking LED lights. The LEDs are switched on and off in succession and this orchestrated blinking of lights creates the impression of motion. The notable difference between these two viewpoints is that in the physical view of motion, the prior states have ceased to exist but in the LED simulation the prior states still exist as LEDs that have been switched off. While these LEDs are not visible, they exist in some sense. Similarly, in the semantic view of change, the past states exist as ideas although they cannot be observed. The past does not become the present. Rather, the past disappears and the present appears. All states of matter exist simultaneously as ideas and only a subset of these ideas are manifest at one time. Furthermore, in the physical view of change, the objects in different states are the same *type* but in the semantic view of change, the objects in different states are also different types. The motion of an object is thus a type evolution.

As representations of ideas, each of these states are symbols and the evolution of the object is a succession of symbols, quite like the sequence of words in a book. But this raises a problem: traditionally we can predict the evolution of word-order only if we know the meaning in the book. However, we don't know the meaning! How can science then predict the evolution of the universe?

The sequence of symbols can be predicted in science but it requires us to reconceive the nature of scientific laws. In current science, space-time is flat and infinite and laws of nature predict the change in object states based on object properties *inside* space-time. In a semantic universe, object states are simply the structure of space-time itself—i.e. the manner in which space and time are divided and organized hierarchically. There is no difference between matter and space-time in the semantic view and therefore to describe the evolution of the universe the scientific theory would have to describe the evolution of space-time structure itself. The classical causal paradigm where we describe the evolution of objects *in* space-time now collapses; we need to describe the evolution of space-time and this evolution is not about an object *inside* anything. However, this shouldn't be a cause for concern if space-time structure is closed and hierarchical because this structure will itself entail a universe that cyclically redistributes matter or energy.

The universe is, in a sense, a gigantic vibration although the vibration is not an object (a string or membrane) *in* space-time. Rather,

space-time itself is the vibration. Such a universe would expand and collapse not because of matter and force in it, but because of the structure
of space-time itself. Such a universe will begin in an elementary concept
or meme which is then divided into many more parts under the effect of
time. Each of these parts will then further divide into more parts, creating ever more diversity. However, at some point in time, this diversification would reverse and the universe would merge back into the original
meme, only to be created at some later point in time. The instances
when time divides primordial ideas into detailed ideas amount to the
addition of information which creates diversity. Similarly, the instances
when time collapses detailed ideas into primordial ideas amount to
removal of information which destroys diversity. The cyclic nature of
time would therefore create a hierarchical space and then destroy it.

The only interesting question left at this time would be: Why is time
cyclic and why is the origin of the universe an idea? That is, if all ideas
are eternally possible, why aren't these ideas always manifest? Why
does time create and destroy diversity? This would, however, not be a
question that science can tackle directly. Science can tackle questions
of 'how' rather than 'why' especially when 'why' represents the quest
for a *justification* rather than an *explanation*.

This view of nature raises some moral problems of choice. If nature
is entirely determined then how do we have free will? This problem
requires a more exhaustive discussion than is possible in a few short
passages here and I will therefore not attempt to elaborate on that
problem here. I have separately discussed[40] the problem of free will
based on Benjamin Libet's experiments on the brain which show
that we don't have free will although we have a free won't. Thus, for
instance, conscious choices cannot produce anything new that matter itself could not have produced although these choices can accept
something else that nature automatically produces. In a deterministic
nature, conscious choices don't change what occurs in the universe
but they can determine which aspects of the universe's events the
observer participates in. In a sense, we have the ability to select which
roles in the cosmic drama we participate in, although the drama itself
is predetermined and cannot be changed. The nature of conscious
choice itself has to be understood in a new way in light of the problem
of material determinism.

Epilogue

Much of the criticism of evolutionary theory today rests on the premise that the theory paints a deterministic picture of nature, leaving no room for free will, upon which moral theories can rest. My criticism of the theory, on the other hand, is that it is not deterministic enough. Indeed, my criticism of current science as a whole is that it is not deterministic enough. As I outlined in the book, there are problems of indeterminism, incompleteness, incomputability, and irreversibility in all areas of science, including mathematics, physics, and computing. Biology just inherits these problems although biologists don't seem to recognize their existence and tend to revert back to Newtonian ideas about matter, motion, and causal determinism in painting their evolutionary paradigm, which simply does not exist in current science. Only when one realizes the extent of the problem and the possible alternatives that exist to address these issues can a serious conversation about evolution occur.

One must, however, be careful in distinguishing the issue of determinism from that of reductionism. Reductionism, in the context of biology, refers to the reduction of biological function and mental states to the properties of independent physical objects, as they were conceived in classical physics. Newtonian particles possess properties that are unaffected by the properties of other objects. For instance, the mass of an object in classical physics is independent of the mass of other objects. The world of such particles cannot represent qualities, intentionality, and contextuality, and therefore the mind's capabilities cannot be explained using a theory of such particles. Furthermore, the world of Newtonian particles cannot explain biological function as the coordinated working of particle collections, because a set of independent particles is never coordinated (we saw earlier how order emerges when boundary conditions are applied, and the order is therefore a byproduct of these conditions).

Evolutionists have postulated random mutation and natural selection as the mechanism for creating order in nature. This idea has no grounding in physics because all physical theories are indeterministic with regard to matter distribution. There are infinitely many different ways in which matter (or energy) can be divided into peg-hole combinations, and there is no physical principle by which one of these combinations can be selected. In fact, as we saw earlier, indeterminism arises as a consequence of reduction. The attempt to think about matter as independent particles leads to the problem of probability in quantum theory, irreversibility in statistical mechanics, and indeterminism in relativity. The idea that numbers are produced from independent object collections leads to Gödel's incompleteness and the idea that a program is an arbitrary sequence of independent instructions leads to the insolvability of the Halting Problem. All theories in modern science begin in the idea of independent objects, and this notion of reduction that aims to reduce all observations to properties of independent objects is known to fail in all areas of science not because we cannot postulate this reduction but because if we do then the theory is incomplete.

The claim therefore that evolution is wrong because it paints a deterministic picture of nature is false. Evolution is wrong because it tries to reduce biological function and mind to material objects and that results in an indeterministic description. The problem is not that we are headed towards determinism but that we don't have a sufficiently powerful deterministic theory that can predict. Furthermore, since this failure is a fundamental theoretical problem in computing, mathematics, and physics, it cannot be solved by the discovery of more empirical minutiae about brain and biology.

Does this failure imply a failure of reduction whereby we give up the attempt to reduce the complex to the simple? Or does it entail the need to think about reduction in a new way? I believe that science needs a new way of carrying out reduction. In this scheme, material objects are symbols of meanings and they are produced from meanings. Therefore, we do not reduce nature to elementary independent objects. Rather, we reduce nature to basic ideas.

The universe of independent objects exists in a flat and infinitely extended space-time. The universe of symbols and ideas exists in a hierarchical and closed space-time. There are many everyday ana-logues of this hierarchical notion of space-time ranging from postal addresses and clock times to Internet addresses and family names. Biologists too speak about the diversification of species from a single species, constructing a tree of phylogenetic diversification. The differ-ence between physical and semantic reduction is that physical reduc-tion treats all the diversified objects as being mutually independent while semantic reduction treats them as conceptually related by the fact that they are diversified from a common meme.

A Terrier and a Doberman are therefore related by the fact that they are both modifications of a dog; a dog and a cat are related by the fact that they are both mammals and so forth. The phylogenetic hierarchy that evolutionists draw in biology is not just a conceptual or historical map of how diversity was created in the universe. It is rather also a map of how species are currently related in nature. This relationship entails that they cannot evolve independently. If nature is idea-like and dogs and cats are variations of the idea of mammals then both dogs and cats must evolve simultaneously when the idea of mammals evolves. Now, we cannot describe the independent evolution of all the parts in the universe. We must rather describe the collective evolution of the universe as a whole.

The evolution of matter—and this evolution includes the appear-ance and disappearance of biological species—should be modeled around the insights about the evolution of ideas, societies, cultures, and history itself. As the popular saying enunciates, history repeats itself. The same ideas recur again and again although the laws of this ideological evolution are not well-known. When an idea has dis-appeared from a society or culture, it still exists as a possibility that can be spoken of although it will likely not be practically realized. In that sense, all species are eternal; they appear and disappear with the passing of time, thereby creating new types of bodies. These ideas are related to other ideas, and hence the species are related to other spe-cies, in a phylogenetic tree, just as all ideas are produced from some primordial ideas in a hierarchical fashion. The tree itself is finite but not all branches of the tree are simultaneously manifest as material objects, living beings, and their ecosystems.

Endnotes

THE BIG PICTURE OF EVOLUTION

1   By environmental mutation I here mean the changes to other species or environmental factors such as weather, soil, water, etc.

2   When matter is reduced to meanings, the conservation of matter is simply the fact that there is matter and anti-matter and the total amount of matter or energy is always zero. The universe as a whole is therefore created from a state of material nothingness, but how this nothingness becomes something requires a semantic understanding.

META-BIOLOGICAL CONSIDERATIONS

3   For the case of biological diversity it seems correct to assume that the probabilities should be finite, although in principle it is possible to assume that the possibility set over which these probabilities have to be defined is infinite.

4   As long as an observation is not made, the quantum formalism is linear. The measurement process is outside the present quantum formalism, so whether the theory that predicts the outcomes of measurements would also be linear is not yet known. However, I will later show that these predictions can be linear in a theory that treats quantum objects as symbols of meanings. This linear theory would however not be reversible in time, since time will also get a direction.

5   Prior to the advent of Newton's mechanics, there were profound and paralyzing questions about the origin of the universe and motion. For motion to originate from a motionless state there had to be an Unmoved Mover, which was supposed to be God. Newton changed the problem from

the origin of motion to changes to motion. His first law of motion postulates that an object once set into motion would continue to be in motion, and this motion was called the property of *momentum*. Newton's theory required an initial choice of position and momentum after which everything could be predicted deterministically.

6   It is possible to further deconstruct this notion of an observer into parts which produce and consume meanings and which own the meaning. The ownership of meaning is the intentionality of meaning by which we suppose that some meaning 'belongs' to an object, because it was created by that object. That deconstruction, however, is not essential for our present discussions, although the distinction between the producer-consumer and owner is pertinent to a more detailed understanding of the mechanisms by which meanings are produced and intentionally owned by observers.

7   Cristian S. Calude, Michael A. Stay, "Most programs stop quickly or never halt," Advances in Applied Mathematics 40 (2008) 295–308.

8   I don't mean to imply that the brain is limited to just meanings. There are intentions, emotions, sensations, and free will which are equally problematic. I just pick 'meaning' because it appears to be the least problematic of all from a standpoint of physical theories. For instance, a computer scientist who hopes to replicate the mind's abilities in a machine does not hope to replicate emotion, intention, sensation, or free will, although he or she hopes that meanings can be encoded in a machine's physical states. Therefore, rather than muddle the discussion with questions about emotion, intention, sensation, and free will, I will restrict myself to questions of meaning which seem far more tractable from the standpoint of where current physics and biological thinking stands.

9   This fact is closely related to the problem in quantum theory, which predicts probabilities of quantum states but cannot predict the next quantum state. If the quantum world is a world of symbols, then we can see how measurement of physical properties will give us probabilistic predictions—since the symbol of meaning can be denoted by its physical properties, i.e., shape or vibration—but cannot help us determine the meaning and hence the symbol order. This topic is dealt with at length in my book *Quantum Meaning*.

10 My earlier book *Gödel's Mistake: The Role of Meaning in Mathematics* discusses the genesis of several logical, mathematical, computational, and philosophical paradoxes and traces them to the inability to hold categorical distinctions in a formal language. These distinctions exist in ordinary language and when language categories are brought into a formal language such as mathematics or logic, various types of contradictions are created.

11 This is also called Gödel numbering which represents the Arithmetization of Syntax rather than Semantics.

12 For details of this argument and how it relates to Turing's Halting Problem, refer to *Gödel's Mistake*.

13 Common illustrations of this fact are seen in the everyday world. For instance, you can a use a knife for many different jobs such as cutting, piercing, weighing, etc. Likewise, you perform a particular job such as cutting using several different objects like knife, blade, saw, hammer, etc.

14 In game theory, the Nash equilibrium is a state of a non-cooperative game, in which no player can improve their payoff by unilaterally changing their strategy.

MATTER AND MIND

15 Manus, Margaret (1999), Gods of the Word: Archetypes in the Consonants, Truman State University Press.

16 Ramachandran, V. S. and Hubbard, E. M. (2001b). "Synaesthesia: A window into perception, thought and language." Journal of Consciousness Studies 8 (12): 3–34.

17 My earlier book *Quantum Meaning: A Semantic Interpretation of Quantum Theory* describes the application of semantic ideas to quantum reality. The book shows that the quantum problems of uncertainty, indeterminism, statistics and non-locality can be understood if matter is viewed as symbols rather than things. If matter encodes meanings then there are additional facts about nature that current science does not capture. The interpretation makes it possible to see how semantic facts about objects will lead to new predictions.

18 This is made possible through the Ensemble Interpretation that Einstein used to explain the quantum problem. Einstein claimed that quanta are individual particles and they arrive one by one at the detectors. However the order in which they arrive cannot be predicted.

19 I will later discuss at length the problems of incompleteness in mathematics and computing which also arise because mathematics cannot incorporate meanings into a syntactical language.

20 Physical theories that employ the idea of hidden dimensions suppose that our observations are themselves incomplete and this incompleteness will be resolved by postulating higher dimensions and hidden properties that are not normally available to perception may be available as science probes these higher dimensions. In these theories the attempt is to try to complete a physical theory by advancing new types of objects that exist in hidden dimensions.

21 The term 'empirical' here denotes conventional forms of empiricism such as the perception through eyes, nose, ears, skin, and tongue. The empirical is indeed physical, when the definition of empiricism is broadened to include other more abstract forms of reality.

22 Dalela, A (2014). *Godel's Mistake: The Role of Meaning in Mathematics*. Shabda Press.

23 The conflict between quantities and types is seen in mathematics where it appears as logical paradoxes. I will later show how mathematical paradoxes illustrate that a description of types cannot exist if mathematics treats symbols as things rather than meanings.

24 It should be remembered that the meanings in the DNA are not entirely in the DNA itself. Rather, the meanings are an outcome of the DNA structure in conjunction with the other molecules or the 'environment' of the DNA. This environment can modify the meanings of the DNA, as research on epigenetics suggests.

25 Darwin, Charles, *The Origin of Species*, chapter V, Laws of Variation.

AN ALTERNATIVE EVOLUTIONARY THEORY

26 http://en.wikipedia.org/wiki/Named_data_networking

27 Mandelbrot, Benoît B. (1983). The fractal geometry of nature. Macmillan.

28 The following articles discuss variations in radioactive decay.

D. Javorsek II, P.A. Sturrock, R.N. Lasenby, A.N. Lasenby, J.B. Buncher, E. Fischbach, J.T. Gruenwald, A.W. Hoft, T.J. Horan, J.H. Jenkins, J.L. Kerford, R.H. Lee, A. Longman, J.J. Mattes, B.L. Morreale, D.B. Morris, R.N. Mudry, J.R. Newport, D. O'Keefe, M.A. Petrelli, M.A. Silver, C.A. Stewart, B. Terry. "Power spectrum analyses of nuclear decay rates". Astroparticle Physics. Volume 34, Issue 3, October 2010, Pages 173–178. http://www.sciencedirect.com/science/article/pii/S0927650510001234.

P.A. Sturrock, G. Steinitz, E. Fischbach, D. Javorsek II, J.H. Jenkins. "Analysis of gamma radiation from a radon source: Indications of a solar influence". Astroparticle Physics, Volume 36, Issue 1, August 2012, Pages 18–25. http://www.sciencedirect.com/science/article/pii/S0927650512000928.

Jere H. Jenkins, Kevin R. Herminghuysen, Thomas E. Blue, Ephraim Fischbach, Daniel Javorsek II, Andrew C. Kauffman, Daniel W. Mundy, Peter A. Sturrock, Joseph W. Talnag. "Additional experimental evidence for a solar influence on nuclear decay rates". Astroparticle Physics, Volume 37, September 2012, Pages 81–88. http://www.sciencedirect.com/science/article/pii/S0927650512001442.

29 D. Lal, A.J.T. Jull, D. Pollard, L. Vacher; Jull; Pollard; Vacher (2005). "Evidence for large century time-scale changes in solar activity in the past 32 Kyr, based on in-situ cosmogenic 14C in ice at Summit, Greenland". Earth and Planetary Science Letters 234 (3–4): 335–249. http://www.sciencedirect.com/science/article/pii/S0012821X05001135.

30 Henrik Svensmark, Eigil Friis-Christensen. "Variation of cosmic ray flux and global cloud coverage—a missing link in solar-climate relationships". Journal of Atmospheric and Solar-Terrestrial Physics, Volume 59, Issue 11, July 1997, Pages 1225–1232. http://www.sciencedirect.com/science/

article/pii/S1364682697000011.

31 Sheldrake, Rupert (2012). The Science Delusion. Coronet.

32 Dalela, Ashish (2014). Quantum Meaning: A Semantic Interpretation of Quantum Theory. Shabda Press.

COMPARATIVE ANALYSIS

33 Behe, Michael (1998). Darwin's Black Box: The Biochemical Challenge to Evolution. Free Press.

34 Kuhn, Thomas (1979), "The Essential Tension: Selected Studies in Scientific Tradition and Change," University Of Chicago Press.

35 Stephen Jay Gould (2007), "The Richness of Life: The Essential Stephen Jay Gould," W. W. Norton & Company.

36 Elgar G, Vavouri T (July 2008). "Tuning in to the signals: noncoding sequence conservation in vertebrate genomes". Trends Genet. 24 (7): 344–52.

37 Peter Andolfatto, "Adaptive evolution of non-coding DNA in Drosophila," Nature 437, 1149-1152 (20 October 2005).

38 This is of course not the only possible scheme of drawing a parallel between computer programs and biological information. However, it is used here to illustrate some basic concepts about the interdependence between the three types of information.

39 Shannon, C.E. (1948), "A Mathematical Theory of Communication," Bell System Technical Journal, 27, pp. 379–423 & 623–656, July & October, 1948.

40 Ashish Dalela (2014), "Is the Apple Really Red? 10 Essays on Science and Religion," Shabda Press.

Acknowledgements

The inspiration behind this book lies in the writings of His Divine Grace A.C. Bhaktivedanta Swami Prabhupāda. He spoke about matter and science with as much ease as he did about soul and God. From his work I first came to believe that there is indeed an alternative way of looking at the material world, different from how it is described in modern science. I am deeply indebted to him in more ways than I can express here in a few words.

The book in the current form would have been impossible without the tireless efforts of Ciprian Begu. He has been my friend and partner in bringing this to life. He read through drafts, edited, did the layout and helped with the cover design. He has tried to teach me the nuances of English grammar, although I haven't been a good student. He figured out all the nits on publishing—something that I did not have the time, energy or the inclination for.

I would like to thank my long-time friend Rukesh Patel. His exuberance, encouragement, patient hearing and drive have helped me in innumerable ways. We have laughed so much together—often at our own stupidity and ignorance— that simply thinking of him makes me smile.

I would like to thank Prof. Pinaki Gupta-Bhaya, my professor and supervisor at IIT Kanpur, who showed me the beauty and excitement of science. From him I learnt that it was not important to know everything, as long as you know where to find it. Looking at his breadth and depth, I came to believe it was possible to step out of the parochial boundaries in science.

My immense gratitude also goes to my parents, who taught me honesty, hard work and simplicity. They gave me the values and upbringing for which I am deeply indebted. My heart also reaches out to my daughter, whose affection and kindness inspires me everyday to become a better person. My wife has been the leveling force in my life. She keeps me grounded to reality, distills complex problems into a succinct bottom-line, and manages the relationships that I would not.

And finally a big thank you to all my readers who have, over the years, written (and continue to write) showing a deep sense of excitement about these books. Their encouragement continues to instill confidence in me that there is a need for these types of books.

My Story

I have always had a great curiosity for the inner workings of nature, the mysteries of the human mind and the origins of the universe. This naturally drew me towards pure sciences. My father, a more practical man, saw this interest as pointless; he was upset when I chose a 5-year program in Chemistry at IIT Kanpur rather than one of the engineering programs, which stood to offer me a better career.

When I started at IIT Kanpur, I believed that my long-held curiosities about the inner workings of nature would be satisfied by an understanding of science. But as I scraped through the coursework and scoured through nearly every section of IIT's extensive library looking for answers, I found that, contrary to my belief, many fundamental and important questions in science remained unanswered. That prompted me to turn towards other departments—since chemistry pointed towards physics which in turn pointed towards mathematics, it seemed that the answers lay elsewhere. However, as I sat through courses offered by other departments—mathematics, physics and philosophy—my worst fears began to materialize: I realized that the problems required discarding many fundamental assumptions in science.

That started me on a journey into the search for alternatives, which has now been spanning 20 years. It was not uncommon in India for students in elite institutions to spend a lot of time discussing philosophy, although often in a tongue-in-cheek manner. My intentions were more serious.

I studied Western philosophy—both classical and modern—as well as Eastern ideas (such as Zen and Taoism) before turning towards Vedic philosophy. I was primarily interested in the nature of matter, the mind and the universe and only Vedic philosophy seemed to offer the kind of synthetic detail I was looking for. I suspected that if the ideas of reincarnation, soul and God in Vedic philosophy were connected to a different view of matter, mind and the universe, then I might actually find an alternative view that could solve the problems in modern science.

At the end of my 5 years at IIT Kanpur, I knew I wanted to pursue the alternative, but I wasn't quite clear how that could work.

I anticipated the pursuit of an alternative in mainstream academia to be very hard. The development of the alternative would frequently run into opposition, and would not fit into the publish-and-tenure practices. A reasonable understanding of ideas often requires longer discussions which may not fit into 3000-word papers. Alternatives often require stepping outside the parochial boundaries of a single field and the journals that accepted such multi-disciplinary articles did not exist at that time—they are more common now.

I therefore faced a difficult choice—pursue a mainstream academic career and defer the search for alternatives until I had established a reputation through conventional means, or pursue a non-academic career to finance my interest in academic alternatives. I chose to separate academics from profession. It was a risky proposition when I started, but in hindsight I think it has worked better than I initially imagined. This and my other books are byproducts of my search for answers to the problems in science, outside mainstream academia.

My career is that of a computer engineer and I have worked for over 17 years in multi-national corporations on telecommunications, wireless and networking technologies. I have co-authored 10 patents and presented at many conferences. I live in Bangalore, India, with my wife and 10-year old daughter.

Connect with Me

Has this book raised your interest? You can connect to my blog or get involved in discussions on www.ashishdalela.com. For any questions or comments please e-mail me at adalela@shabdapress.net.

Other Books by Ashish Dalela

Moral Materialism
A Semantic Theory of Ethical Naturalism

Modern science describes the physical effects of material causes, but not the moral consequences of conscious choices. Is nature merely a rational place, or is it also a moral place? The question of morality has always been important for economists, sociologists, political theorists, and lawmakers. However, it has had almost no impact on the understanding of material nature in science.

This book argues that the questions of morality can be connected to natural law in science when science is revised to describe nature as meaningful symbols rather than as meaningless things. The revision, of course, is entailed not just by issues of morality but also due to profound unsolved problems of incompleteness, indeterminism, irreversibility and incomputability in physics, mathematics, and computing theory. This book shows how the two kinds of problems are deeply connected.

The book argues that the lawfulness in nature is different from that presented in current science. Nature comprises not just *things* but also our *theories* about those things. The world of things is determined but the world of theories is not—our theories represent our free will, and the interaction between free will and matter now has a causal consequence in the evolution of scientific theories.

The moral consequences of free will represent the ideological evolution of the observer, and the correct theory represents the freedom from this evolution. Free will is therefore not the choice of arbitrary and false theories; free will is the choice of the correct theory. Once the correct theory is chosen, the observer is free of natural laws, since all phenomena are consistent with the correct theory.

Uncommon Wisdom
Fault Lines in the Foundations of Atheism

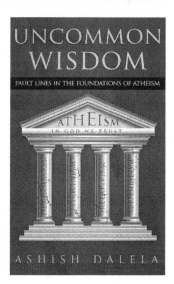

The rise of militant atheism has brought to fore some fundamental issues in our conventional understanding of religion. However, because it offers science as an alternative to religion, militant atheism also exposes to scrutiny the fundamental problems of incompleteness in current science.

The book traces the problem of incompleteness in current science to the problem of universals that began in Greek philosophy and despite many attempts to reduce ideas to matter, the problem remains unsolved. The book shows how the problem of meaning appears over and over in all of modern science, rendering all current fields—physics, mathematics, computing, and biology included—incomplete. The book also presents a solution to this problem describing why nature is not just material objects that we can perceive, but also a hierarchy of abstract ideas that can only be conceived. These hierarchically 'deeper' ideas necessitate deeper forms of perception, even to complete material knowledge.

The book uses this background to critique the foundations of atheism and shows why many of its current ideas—reductionism, materialism, determinism, evolutionism, and relativism—are simply false. It presents a radical understanding of religion, borrowing from Vedic philosophy, in which God is the most primordial idea from which all other ideas are produced through refinement. The key ideological shift necessary for this view of religion is the notion that material objects, too, are ideas. However, that shift does not depend on religion, since its implications can be known scientifically.

The conflict between religion and science, in this view, is based on a flawed understanding of how reason and experiment are used to acquire knowledge. The book describes how reason and experiment can be used in two ways—discovery and verification—and while the nature of truth can never be *discovered* by reason and experiment, it can be *verified* in this way. This results in an epistemology in which truth is discovered via faith, but it is verified by reason and experiment.

Quantum Meaning
A Semantic Interpretation of Quantum Theory

The problems of indeterminism, uncertainty and statistics in quantum theory are legend and have spawned a wide-variety of interpretations none too satisfactory. The key issue of dissatisfaction is the conflict between the microscopic and macroscopic worlds: How does a classically certain world emerge from a world of uncertainty and probability?

This book presents a Semantic Interpretation of Quantum Theory in which atomic objects are treated as symbols of meaning. The book shows that quantum problems of uncertainty, indeterminism and statistics arise when we try to describe meaningful symbols as objects without meaning.

A symbol is also an object, although an object is not necessarily a symbol. The same object can denote many meanings in different contexts, and if we reduce symbols to objects, it naturally results in incompleteness.

This book argues that the current quantum theory is not a final theory of reality. Rather, the theory can be replaced by a better theory in which objects are treated as symbols, because this approach is free of indeterminism and statistics.

The Semantic Interpretation makes it possible to formulate new laws of nature, which can be empirically confirmed. These laws will predict the order amongst symbols, similar to the notes in a musical composition or words in a book.

Gödel's Mistake
The Role of Meaning in Mathematics

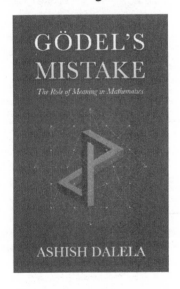

Mathematics is the queen of sciences but problems of incompleteness and incomputability in mathematics have raised serious questions about whether it can indeed be used to describe nature's entire splendor. Proofs that demonstrate the incompleteness and incomputability are respectively called Gödel's Incompleteness and Turing's Halting Problem.

This book connects Gödel's and Turing's theorems to the question of meaning and shows that these proofs rest on what philosophers call category mistakes. Ordinary language contains many categories - such as names, concepts, things, programs, algorithms, problems, etc. but mathematics and computing theory do not. A thing can denote many concepts and vice versa. Similarly, a program can solve many problems, and vice versa. A category mistake arises when we reduce one category to another, and this leads to logical paradoxes because these categories are not mutually reducible.

The book shows that the solution to category mistakes requires a new approach in which numbers are treated as types rather than quantities. This is called Type Number Theory (TNT) in the book. TNT requires a hierarchical theory of space and time, because it is through a hierarchical embedding that objects become symbols of meanings.

Hierarchical notions of space and time are well-known; for instance postal addresses and clock times are hierarchical. A formal theory of hierarchical space-time will also be a theory of symbols and will address problems of incomputability in computing and incompleteness in mathematics.

Sāṅkhya and Science
Applications of Vedic Philosophy to Modern Science

ASHISH DALELA

Since the time of Descartes, science has kept questions of mind and meaning outside science, and in recent times materialists aim to reduce mind and meaning to matter. Both approaches have failed. There are problems of meaning in mathematics, computing, physics, biology and neuroscience. A new view of nature is needed, one that integrates matter and meaning more directly.

The Vedic theory of matter—called Sāṅkhya—shows a path fruitful to the resolution of modern scientific problems. In Sāṅkhya, material objects are created when the mind transfers meanings or information into space-time. These objects are not meaningless things, but symbols with meanings.

The book shows how a symbolic view of nature can be used to solve the problems of incompleteness and indeterminism in atomic theory, chemistry, biology, mathematics and computing. In the process, the book builds a new foundation for science, based on a semantic or symbolic view of nature.

The book will be of interest to both scientists and philosophers, especially those looking to integrate mind and matter without stepping outside the rational-empirical approach to science.

Is the Apple Really Red?
10 Essays on Science and Religion

Conventional wisdom on science and religion says the former is based on experiment and reason, while the latter is based on faith and belief.

Is the Apple Really Red? discusses how the notions of soul, morality and afterlife in religion can be scientific. But for this to be possible, a new science that studies meanings instead of objects is needed. The clash of ideologies between science and religion—this book argues—is based on an incorrect understanding of matter, disconnected from consciousness, and an incorrect notion of God, disconnected from matter, space and time.

A revision of the current views on religion and science is needed, not only to settle the conflict but also to deepen our understanding of matter (and its relation to consciousness) and God (and His relation to matter, space and time)

Written for the layperson, in 10 essays, the book delineates the Vedic view of matter, God, soul, morality, space and time. The author shows how the existence of the soul and God implies a new view of matter, space and time which is empirical and can be used to form new scientific theories.

Such theories will not only change our understanding of matter but will also change our outlook on religion. Readers interested in the science and religion debate will benefit significantly from the viewpoint described in *Is the Apple Really Red?*

Six Causes
The Vedic Theory of Creation

In Vedic philosophy, creation is modeled as the creative activity of consciousness. Six Causes shows us how the universe's creation can be understood based on insights about our own everyday creative activities.

The nature of material objects when they are created by consciousness is different than when they are independent of consciousness. Six Causes discusses this difference. Essentially, objects in the Vedic view are symbols of meaning originating in consciousness rather than meaningless things.

Different aspects of conscious experience—and the different roles they play in the creation—are called the six causes.

Presented in lay person's language, and written for those who don't have any background in Vedic philosophy, this book will allow you to truly understand the intricacies in Vedic texts.

In the process, you will also see many common misconceptions about Vedic philosophy being overturned through a deeper understanding of not just soul, God, reincarnation and karma but also matter, senses, mind, intelligence, ego and the unconscious.

Did You Like Signs of Life?

If you enjoyed this book or found it insightful I would be grateful if you would post a short review on Amazon. Your feedback will allow other readers to discover the book, and can help me improve the future editions. If you'd like to leave a review then go to the website below, click on the customer reviews and then write your own.

http://www.ashishdalela.com/review-amazon-sol

Find Out in Advance When My Next Book Is Out

I'm always working on the next book. You can get a publication alert by signing up to my mailing list on www.ashishdalela.com. Moreover, If you want to receive advance copies of my upcoming books for review, please let me know at adalela@shabdapress.net.